BIOSYNTHESIS OF THE MAJOR CROP PRODUCTS

Wiley Biotechnology Series

This series is designed to give undergraduates and practising scientists access to the many related disciplines in this fast developing area. It provides understanding both of the basic principles and of the industrial applications of biotechnology. By covering individual subjects in separate volumes a thorough and straightforward introduction to each field is provided for people of differing backgrounds.

Published Titles

Series Editors

Professor J. A. Bryant, *Department of Biology, Exeter University, UK*
Professor J. F. Kennedy, *Department of Chemistry, University of Birmingham, UK*

BIOSYNTHESIS OF THE MAJOR CROP PRODUCTS

The biochemistry, cell physiology and molecular biology
involved in the synthesis by crop plants of sucrose, fructan,
starch, cellulose, oil, rubber and protein

Philip John

Department of Agricultural Botany
University of Reading, UK

JOHN WILEY & SONS
Chichester · New York · Brisbane · Toronto · Singapore

Published 1992 by John Wiley & Sons Ltd,
Baffins Lane, Chichester
West Sussex PO19 1UD, England

Other Wiley Editorial Offices

John Wiley & Sons, Inc., 605 Third Avenue,
New York, NY 10158–0012, USA

Jacaranda Wiley Ltd, G.P.O. Box 859, Brisbane
Queensland 4001, Australia

John Wiley & Sons (Canada) Ltd, 22 Worcester Road
Rexdale, Ontario M9W 1L1, Canada

John Wiley & Sons (SEA) Pte Ltd, 37 Jalan Pemimpin #05–04,
Block B, Union Industrial Building, Singapore 2057

Library of Congress Cataloging-in-Publication Data

John, Philip.
  Biosynthesis of the major crop products : the biochemistry, cell
physiology, and molecular biology involved in the synthesis by crop
plants of sucrose, fructan, starch, cellulose, oil, rubber, and
protein / Philip John.
     p.   cm.—(Biotechnology series)
  Includes bibliographical references and index.
  ISBN 0 471 93585 9 (cloth)   0 471 93816 5 (paper)
     1. Crops—Physiology.   2. Plant products—Synthesis.   3. Plant
biotechnology.   I. Title.   II. Series.
  SB112.5.J64   1992
  582.13′041929—dc20   92-18692
                                                                  CIP

British Library Cataloguing in Publication Data

A catalogue record for this book is available from the British Library

ISBN 0 471 93585 9 (cloth)   0 471 93816 5 (paper)

Typeset in 10/11½pt Times by Acorn Bookwork, Salisbury, Wiltshire
Printed and bound in Great Britain by Biddles Ltd, Guildford, Surrey

To
Luciana

Contents

Preface and Acknowledgements

This book describes how crop plants synthesise and accumulate sucrose, fructan, starch, cellulose, oils, rubber and protein. Produced on an enormous scale, these are the bulk chemicals of agricultural production. Sugar, starch, oils and protein together form the basis of our diet. As commodities traded around the world these crop products are second in value only to petroleum oil. Yet despite their extraordinary importance in agriculture, nutrition and world trade, relatively little attention is paid to their biosynthesis by student textbooks.

The present book is based on a course of lectures that has been given to undergraduates in their Final Year of Crop Science, and to postgraduates studying for a MSc in Crop Physiology. I have assumed an elementary knowledge of biochemistry, so that the reader should be familiar with the fundamentals of carbohydrate metabolism, protein synthesis, and the structure and physiology of plant cells. The book is therefore suitable for undergraduates in their second and third years studying for degrees in the Plant and Agricultural Sciences, and those students of Biotechnology and Biochemistry who have an interest in the productivity of crops.

The eminent Jack Preiss has often commented that 'There is more to starch synthesis than the polymerisation of glucose units'. For a start, there is the production of a genetically fixed ratio of the two polymeric components, amylose and amylopectin, and then there is their ordered assembly into the distinctive structure of the starch grain. In this book I have applied Professor Preiss's observation to all the crop products covered, so that I have not been limited by the metabolic pathway leading to a chemically defined compound, but for each product I have described—as far as our present knowledge permits—its assembly in the chemical and physical form in which it is accumulated and finally harvested.

I have not attempted to treat the biosynthesis of different crop products in a uniform way. Rather have I tried in each case to reflect our present knowledge, with all its unevenness. Thus, for example, we can now direct the synthesis of a modified storage protein in a transgenic plant, but the elaboration of the structure of a starch grain or of a cellulose fibre remain largely mysterious. Even the mechanism responsible for the accumulation of sucrose in sugar beet and cane is not yet known.

The inclusion of fructan (Chapter 4) may be surprising. Currently it is produced commercially on a tiny scale, and from crops which are not widely grown. Moreover there is much to learn about the biochemistry of its synthesis. However, it is included because it does have great potential as a crop product with applications in the food and non-food industries. Also, an account of fructan biosynthesis illustrates well how a better knowledge of the enzymes involved and the biochemical pathway followed during its synthesis are essential to fully realise the potential of fructan as a significant crop product.

For each product I have indicated where our biochemical knowledge provides a basis for future biotechnological developments. In the earliest times technology was based entirely on natural products, then on coal, now it is based on oil. However, concerns about agricultural overproduction in the Developed World, and the environmental burden of non-biodegradable materials are leading towards a renewed interest in natural products as raw materials for industry. Over the past 20 years the revolutionary discoveries in molecular biology have provided the tools which are making it possible to redesign the products manufactured by our crops. To make effective use of this newly acquired technology we need to know how crop plants make, assemble and store the products they accumulate. The role of this book is to provide that biochemical background.

I am grateful to those authors who sent me reprints and preprints of their papers, and who generously provided illustrative material. I am indebted to Luciana Rosa John and Professor F.R. Whatley FRS for helpful criticisms during the preparation of the manuscript.

Chapter 1

Introduction

Worldwide, the crop products described in this book are produced and traded on a massive scale (Table 1.1). Locally they can dominate an economy, with whole areas being dedicated to the production of a single product: cane sugar in the wet tropics, rubber and oil from the plantations of Malaysia and Indonesia, cellulose in the form of cotton from the Southern States of the USA.

Despite the diversity of the crop products in chemical and physical properties, some generalisations can be made. They are predominantly carbohydrates—sucrose, starch, fructan and cellulose—or hydrocarbons—rubber and the oils (mainly hydrocarbon)—except for the proteins, where nitrogen and sulphur enter into their composition. They are all polymeric, except for sucrose. They are all insoluble, except for sucrose and fructan.

It is in the nature of modern agricultural production that a relatively small number of species are responsible for a large part of the world production of any particular product. Nevertheless, crop plants are far from being unique in producing the particular product. Thus almost the entire agricultural production of sucrose comes from just two crops, sugar cane and sugar beet, yet every green plant can make sucrose. The rubber tree is the sole commercial source of the 4 million tonnes of natural rubber produced commercially every year, even though rubber is synthesised by over 2000 species of plants. The essential characteristic which distinguishes those species exploited as crop plants is their ability to accumulate a useful product in a harvestable form. This ability is accompanied by a combination of other desirable agronomic traits, all of which have been selected for during the evolution of the crop. But without the ability to synthesise, accumulate and store a useful product, the plant would not have been considered for domestication initially.

Table 1.1   World trade in some crop products.

Commodity	Product	Total world exports in 1989 (billion US dollars)
Cane, beet sugar	Sucrose	11.7
Wheat	Starch, protein	18.2
Maize	Starch, protein	10.3
Cotton	Cellulose	8.4
Palm oil	Oil	2.6
Natural rubber	Rubber	4.3
Soybeans	Protein, oil	6.2

(Data from FAO Trade Yearbook, 1989)

When plant physiologists study the assimilation and distribution of carbon in a crop plant, different parts of the plant are identified as either sources or sinks. Whether a particular part of the plant is a sink or a source will change with the development of the crop plant during its life-cycle. For example, the cereal endosperm is a strong sink during grain development, but becomes a source of assimilate for the germinating grain. However, during the synthesis and accumulation of a crop product the principal source of assimilate is represented by the green, photosynthetically active parts of the plant, usually the leaves, while the principal sink is represented by the maturing storage organs. The capacity of these storage organs for growth and development constitutes an important factor in determining the potential yield.

In the evolution of crop plants a variety of tissues and organs have been developed as sites for the deposition of the crop product (Table 1.2). Even when considering a single product variation is evident. For example, other than the ability of their parenchyma to accumulate large amounts of sucrose, there is little in common between the concentric rings of a beet root and the internodes of sugar cane stem. The physical framework in which the crop products are deposited is provided by a pattern of cell division and expansion characteristic for the particular storage organ. The integration of that development with the deposition of the product can also vary from one crop to another. For example, in a potato tuber starch synthesis is accompanied by continued cell division and expansion, while in the cereals starch fills an endosperm in which cell division has ceased.

At the subcellular level, the chloroplast can be identified as the organelle responsible for assimilate production at the physiological source, but no single organelle can be said to be responsible for assimilate synthesis and deposition at the physiological sink (Table 1.3).

The vacuole in a variety of disguises could be said to be responsible for the storage of a number of crop products; sucrose and fructan are the more obvious cases, but oil and protein can also be considered to be deposited in modified vacuoles. Often synthesis occurs in a cell compartment different

Table 1.2 Storage tissues and organs of the major crop products.

Product	Tissue	Organ
Sucrose	Storage parenchyma	Root (beet)
		Stem (cane)
Fructan	Storage parenchyma	Root and stem tubers
Starch	Endosperm	Seeds (cereals)
	Storage parenchyma	Tubers
Cellulose	Epidermal	Seed fibres (cotton)
	Secondary xylem	Stem (wood)
Oils	Cotyledons	Seeds
	Storage parenchyma	Fruit (oil palm)
Rubber	Secondary phloem latex	Stem
Protein	Endosperm	Seeds (cereals)
	Cotyledons	Seeds (legumes)

from that in which the product is stored (Table 1.3); the vacuole is never the site of synthesis, except perhaps in the case of fructan. Consequently transport of the product from one cell compartment to another accompanies the chemical transformations it undergoes during biosynthesis.

In outline the directions taken by the biochemical pathways leading to the biosynthesis of the crop products are as shown in Figure 1.1.

Sucrose is synthesised in the photosynthetic tissue and transferred by the phloem to the storage organs. In sugar cane and sugar beet, sucrose accumulation appears to be simply an accumulative process. In all the other crop products the storage organ is the site of synthesis. Fructan is elaborated directly from sucrose, but usually, sucrose is converted first to an appropriate precursor. For the polymeric carbohydrates, starch and cellulose, glucose units for polymerisation are provided by ADP-glucose and UDP-glucose,

Table 1.3 Intracellular sites of synthesis and deposition of the crop products.

Product	Site of synthesis	Site of deposition
Sucrose	Cytosol	Vacuole
Fructan	Vacuole	Vacuole
Starch	Amyloplast	Amyloplast
Cellulose	Plasma membrane	Cell wall
Oils	Plastids and endoplasmic reticulum	Oil bodies
Rubber	Cytosol	Rubber particles
Protein	Ribosomes on endoplasmic reticulum	Protein bodies

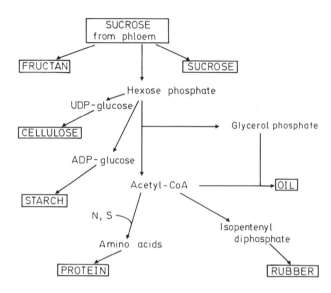

Fig. 1.1 Outline of the flow of assimilate towards the different crop products.

respectively. For rubber and the hydrocarbon chains of plant oils, the key intermediate is acetyl-CoA. To form the triacylglycerols of plant oils the 2-carbon unit of acetyl-CoA is converted into fatty acids 12 to 22 carbons long, which are then esterified to glycerol. In the synthesis of rubber, the acetyl-CoA is first converted to the basic building block for polymerisation, the 5-carbon isopentenyl diphosphate. The biosynthetic pathway to proteins is more complicated than that of any of the other crop products. The 20 protein amino acids are made first, incorporating assimilated nitrogen and sulphur into carbon skeletons derived from intermediates of the tricarboxylic acid cycle and glycolysis. Then polymerisation of the amino acids takes place under the direction of mRNA.

Product synthesis and accumulation would not be accomplished if the enzymes involved were not regulated so that there was a strict coordination of their activities. This enzyme regulation occurs at different levels.

The development of storage organs, like all stages of plant development requires a differential synthesis of gene products in time and in space. This provides the basis for the relatively coarse control which is exerted by the regulation of gene expression, through alterations in the rate of transcription, translation, mRNA processing, and protein turnover. It needs to be borne in mind that fruits and seeds have received more attention from investigators than vegetative structures such as roots, and consequently more is known about the regulation of gene expression for these reproductive structures than for the other storage organs.

A finer metabolic control is exerted by modulating the activity of the pre-existing enzyme without altering its synthesis or degradation. This type of control operates on those enzymes in a pathway that is effectively rate-limiting for assimilate flow through a particular section of the pathway. It enables the biosynthetic pathway to respond quickly to relatively rapid changes in the environment, especially those that might affect the supply of assimilate. A common mechanism of fine control is exerted by the allosteric regulation of enzyme activity, in which metabolites binding to specific sites on the enzyme molecule are able to modulate its activity. In this way the velocity of the particular reaction becomes sensitive to the relative concentration of activating and inhibiting metabolites in the environment of the enzyme. Enzyme activity can also be moderated by a reversible covalent modification of the enzyme molecule, such as phosphorylation brought about by a protein kinase. A number of mechanisms contributing to the fine control of sucrose synthesis in leaf tissues have been elucidated (Chapter 2), but for many of the other crop products little information is available of the mechanisms by which the biosynthesis is regulated either at the level of the gene or at the level of the enzyme.

Chapter 2

Sucrose

INTRODUCTION

Virtually all the carbon assimilated during the photosynthesis of higher plants is channelled into sucrose before the assimilated carbon reaches the non-green parts of the plant. But not only is sucrose the principal form in which organic carbon is transported in higher plants, it also has a role as a storage carbohydrate. Sucrose accumulates transiently in photosynthesising leaves before it is taken away by the phloem, and in storage organs before it is finally converted to starch or oil. In some species sucrose accumulates more permanently and to high concentrations. It is this capacity for a long-term accumulation of sucrose that makes these species important crop plants, and makes sucrose one of the most valuable crop products.

The two main commercial sources of sucrose are the sugar cane and the sugar beet. Cane accounts for about 60 per cent of the world production with beet providing almost all of the remainder, relatively tiny amounts coming from sorghum, sugar palm and maple. The scale of sugar production is enormous. Worldwide about 1000 million tonnes of cane are harvested every year, and the annual production of sugar from cane and beet amounts to over 100 million tonnes. Before sucrose was extracted from these crops, the sucrose in our diet came from fruits and vegetables, and large amounts are still consumed in this way. What chemical and physical features of sucrose make it so important to plants—and to us?

Sucrose is a disaccharide consisting of an α-D-glucopyranoside joined to a β-D-fructofuranose by an α-1,2 linkage (Figure 2.1). It is easily hydrolysed to glucose and fructose by dilute acids and by the enzymes invertase (a β-fructofuranosidase) and sucrase, an α-glucosidase which occurs in the animal intestine. The free energy released on hydrolysis is relatively high (30 kJ/

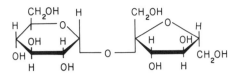

Fig. 2.1 Chemical structure of sucrose.

mol), and comparable to the free energy of hydrolysis of the terminal phosphate of ATP. This feature is often put forward as an important factor in explaining the distinctive role of sucrose in plants. But what may be equally significant is the unreactivity of sucrose compared to glucose and fructose. Unlike its hexose monomers, sucrose is non-reducing and it cannot be directly phosphorylated. Thus sucrose can be viewed as a protected form of glucose and fructose, which does not readily enter metabolism. This unreactivity is more important in plants than in animals, because plant phloem transport is intracellular, while glucose circulates around the animal body in an extracellular medium. In addition sucrose has advantages of being an uncharged molecule, highly soluble, and not inhibitory to metabolic reactions.

The sweetness of sucrose appears to have been appreciated by mankind since the earliest times. Moreover as a crop product, sucrose has been valued for the high yields, the ease with which it can be purified by crystallisation, and the keeping quality of the final product, which has allowed it to be traded widely since the fifteenth century, when Venice controlled the trade.

Most of the sucrose produced in the world goes to satisfy the human desire for sweetness, but in addition, sucrose imparts other desirable qualities to manufactured foods: it improves the physical characteristics of biscuits and cakes by adding bulk, by assisting emulsification, and by imparting plasticity. Its ability to exist in different physical states is important in the manufacture of chocolate and other sweet confectionery. Its great solubility means that solutions of high osmotic potential can readily be obtained, which helps to preserve jams and marmalades in which the low water activity slows the growth of microbes.

Aside from the uses of sucrose in food, sugar cane is the main raw material in the world for producing ethanol from an agricultural source. Large areas of sugar cane are grown in Brazil specifically as an energy crop. The other important feedstocks for producing this bio-ethanol, such as potatoes, cereals and grapes, are grown for other purposes, and only surplus or low quality production is fermented.

The maximum yield of ethanol obtainable from a variety of crops is shown in Figure 2.2. From this comparison it can be seen that the sucrose-accumulating crops compare favourably with those crops that accumulate starch, either in grains or in tubers. The only serious challenge comes from the fructan-accumulating Jerusalem artichoke (see Chapter 4). The amount of ethanol that can be obtained depends on the yield per hectare, and on the content of fermentable carbohydrate. The former depends on climate and

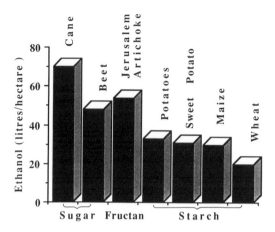

Fig. 2.2   Ethanol yield from carbohydrate-rich crops (data from OECD, 1984).

agronomic practice, while the latter is susceptible to improvements in fermentation technology which would maximise the ethanol production from the biomass produced. The already enormously high yields of sugar cane (over 100 tonnes per hectare) mean that future increases in ethanol production are likely to come from a more effective conversion of the biomass by including the lignocellulose of the stalk fibrous matter, the bagasse.

Brazil has been using ethanol as a fuel since the 1920s, the surplus sugar production being fermented in distilleries near the sugar mills. Now over half the ethanol produced still comes from the sugar solution, called molasses, that remains when crystalline sucrose has been extracted, but about 30 per cent of the area under sugar cane is dedicated to the production of ethanol, and is independent of sugar production. In this area all of the sugar in the crop is used to manufacture ethanol. The ethanol replaces petrol in internal combustion engines; petrol containing 20 per cent anhydrous ethanol performs well without the addition of lead compounds as anti-knock agents.

In addition to its use as a fuel, ethanol is an important industrial solvent, and it is the starting compound for a wide variety of derivatives which are used on a large scale as raw materials in the chemical industry for the production of polyethylene, polystyrene, and polyvinyl acetate. Some countries, such as Brazil and India use fermentation alcohol as a chemical feedstock for their chemicals industry, but most countries use synthetic ethanol derived from petroleum because it is cheaper.

Sucrose is unique among the major crop products in that it is produced in large amounts as a pure chemical of low molecular weight. Unlike cellulose or starch, sucrose is a defined organic chemical, so that its chemical constitution and physical properties do not vary from crop to crop, or from one season to another. This advantage was recognised in the 1940s when research was initiated to investigate the chemistry of sucrose with the aim of developing novel sucrose derivatives, which could open new markets for sucrose. A large number of 'sucrochemicals' were synthesised and numerous applications discovered, but price competition with the large-scale production of the petrochemical industry confines them to specialised markets: sucrose esters of fatty acids are used in the food industry as emulsifiers and stabilising agents, and sucrose ethers are used in the manufacture of polyurethane foams.

BIOSYNTHESIS OF SUCROSE IN LEAVES

Sucrose is the only crop product described in this book which is synthesised in one part of the plant and accumulated in another part. For all the other crop products the organ is which they are deposited is also the organ in which they are formed. Sucrose on the other hand, is synthesised in the leaves and other green parts of the plant, and then transported to the storage parenchyma of sugar cane stems and of the swollen root of sugar beet.

The role of sucrose in the source leaf of sugar beet plants has been described quantitatively by Fondy & Geiger (1982). During the day starch, rather than sucrose, is accumulated by the leaves (Figure 2.3). The accumulated starch is broken down at night, some to furnish respiratory substrates, but most is converted to sucrose which is transported away from the leaf. The large and conspicuous accumulation of starch which occurs during the day should not be allowed to mislead us into viewing starch as quantitatively the more important immediate product of photosynthesis. In fact during the day over twice as much assimilate flows into sucrose synthesis as into starch synthesis (Figure 2.3). The coordination of sucrose and starch metabolism seen in the work of Fondy & Geiger (1982) with sugar beet leaves has been the subject of intensive investigation during the past 10 years. This work has given us a broad understanding at the enzymological level of how assimilate is allocated between starch and sucrose in photosynthetic tissue.

The biosynthesis of sucrose in the photosynthetic parts of plants has been studied in most detail at the biochemical level in a small number of plants notably spinach, wheat and maize, which are easier to maintain in the laboratory than mature beet and cane. Although much of the early work which elucidated the C_4 pathway of photosynthesis was carried out with sugar cane leaves, the synthesis of sucrose in sugar cane and beet has been relatively neglected. We shall therefore be relying for much of the information on leaf sucrose synthesis on studies performed with other plants.

Plants have two enzymes which could potentially be responsible for the biosynthesis of sucrose: sucrose synthase and sucrose phosphate synthase:

Sucrose synthase

$$UDP\text{-glucose} + \text{fructose} \leftrightarrow UDP + \text{sucrose}$$

Sucrose phosphate synthase

$$UDP\text{-glucose} + \text{fructose 6-phosphate} \rightarrow \text{sucrose 6-phosphate} + UDP$$

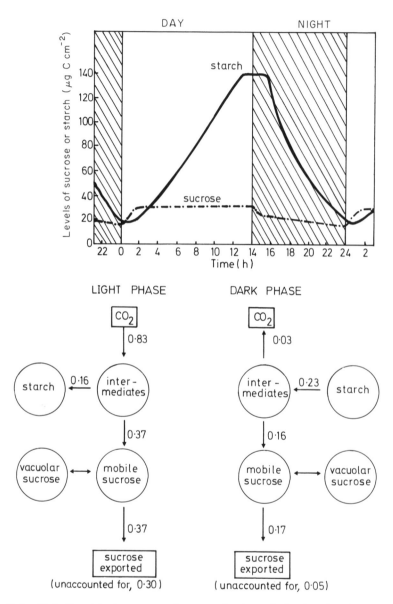

Fig. 2.3 Relationship between photosynthesis and the levels of starch and sucrose in the source leaves of sugar beet plants during a diurnal cycle. The rates of transfer between the pools are given as μg C/cm^2/min. (Simplified from Fondy & Geiger, 1982. Reproduced by kind permission of the authors and of the American Society of Plant Physiologists.)

There is now little doubt, however, that the sucrose phosphate synthase is involved in sucrose synthesis, while the sucrose synthase is involved in the mobilisation of sucrose. The phosphate group of the sucrose phosphate formed by the activity of the sucrose phosphate synthase is removed by sucrose phosphatase. This enzyme is widely distributed in plant tissues, but has a higher activity in tissues actively synthesising sucrose, where its activity is about 10 times that of the sucrose phosphate synthase.

Sucrose phosphatase

Sucrose 6-phosphate + H_2O → sucrose + phosphate

The properties of sucrose synthase and sucrose phosphate synthase are compared in Table 2.1. The evidence that indicates different functions for the two enzymes can be summarised as follows. First, the distribution of sucrose synthase among plant tissues shows it to be ubiquitous, with high activities in non-photosynthetic and storage tissues. By contrast sucrose phosphate synthase activity is largely confined to photosynthetic tissues, with only a low activity detectable in storage and embryonic tissues. As we shall see later, the storage parenchyma of sugar cane is exceptional in having high sucrose phosphate synthase activity.

Second, the sucrose synthase catalyses a reaction which is freely reversible, while the sucrose phosphate synthase catalyses a reaction which is virtually irreversible in the direction of sucrose phosphate synthesis. Clearly the contrasting energetics of the two reactions favour a role in sucrose mobilisation for the sucrose synthase and a synthetic role for the sucrose phosphate synthase.

Third, sucrose synthase shows a broad specificity towards the nucleoside diphosphate. For example, the sucrose synthase from the fibrous roots of sugar beet has a great affinity for ADP than for UDP when catalysing the breakdown of sucrose. This broad specificity would enable the sucrose

Table 2.1 Contrasting properties of sucrose synthase and sucrose phosphate synthase (data from Avigad, 1982).

	Sucrose synthase	Sucrose phosphate synthase
Distribution	Wide	Mainly photosynthetic tissues
Reaction reversibility	Reversible	Irreversible
Substrate specificity	Uses UDP-glucose ADP-glucose GDP-glucose TDP-glucose	Uses UDP-glucose only
Fructose source for sucrose synthesis	Free fructose	Fructose 6-phosphate

synthase to catalyse a variety of nucleoside diphosphate glucosides as substrates for the biosynthesis of different cell components. By contrast the sucrose phosphate synthase is highly specific for UDP-glucose, which appears to be the only glucosyl donor for sucrose synthesis.

Fourth, the affinity of the sucrose synthase for fructose is low, with a K_m of about 5 mM. Although high concentrations of free fructose can be present in some plant organs such as fruits, generally the concentration of fructose in plant tissues is low, and presumably it is confined to either the extracellular cell wall space, or to the vacuole. When fructose is fed to plants, it is rapidly phosphorylated in the cytosol. Thus the substrate for a synthesis of sucrose by sucrose synthase does not appear to be available, by contrast with the fructose 6-phosphate required for the sucrose phosphate synthase, which is readily supplied by cytosolic metabolism. It is worth noting, however, that a role for sucrose synthase in the mobilisation of sucrose, needs to be reconciled with the poor affinity of this enzyme for sucrose. For example with the sucrose synthase extracted from sugar beet roots, the K_m for sucrose is 110 mM. If this is a faithful reflection of the properties of the enzyme *in vivo*, then presumably enzyme activity in the roots is largely governed by the availability of sucrose.

The proper starting point in following the pathway of sucrose biosynthesis is with triose phosphate. This key product of the Calvin cycle in the chloroplast stroma provides a common pool of substrate for starch and sucrose synthesis. The first decisive step in the direction of sucrose synthesis is the efflux of triose phosphate from the chloroplast via the phosphate carrier in a reversible 1:1 exchange for phosphate (Figure 2.4). Triose phosphate in the cytosol is available for the synthesis of fructose 1,6-bisphosphate from the two isomeric forms of triose phosphate, dihydroxyacetone phosphate and phosphoglyceraldehyde. The subsequent metabolic step in the direction of sucrose synthesis is the loss of the phosphate group from the carbon 1 of fructose 1,6-bisphosphate, forming fructose 6-phosphate. The other substrate of the sucrose phosphate synthase, UDP-glucose, is made by the UDP-glucose pyrophosphorylase from glucose 1-phosphate formed by reversible reactions from the fructose 6-phosphate. The pyrophosphate formed was at one time thought to be simply hydrolysed to phosphate by pyrophosphatase activity. However, it is now apparent that pyrophophatase which is active at the pH of the cytosol, (called alkaline pyrophosphatase to distinguish it from an acid pyrophosphatase in the vacuole) is restricted to the chloroplast stroma. One possibility would be for the pyrophosphate to be used to rephosphorylate fructose 6-phosphate to fructose 1,6-bisphosphate by the action of the pyrophosphate fructose 6-phosphate transferase present (Figure 2.4).

Sucrose synthesis in the cytosol of photosynthetic cells is highly regulated so that it is closely coordinated with the fixation of carbon in the chloroplast stroma. During photosynthesis assimilate leaves the chloroplast as triose phosphate, as described above. Continued carbon dioxide fixation requires not only the return of the phosphate, which is ensured by the strict 1:1 exchange mechanism of the phosphate/triose phosphate carrier, but also the

Table 2.2 Enzymes involved in the pathway of sucrose synthesis from triose phosphate in the cytosol of photosynthetic cells.

1. Triose phosphate isomerase
   D-glyceraldehyde 3-phosphate ketol-isomerase   EC 5.3.1.1

   glyceraldehyde 3-phosphate ↔ dihydroxyacetone phosphate

2. Aldolase
   D-fructose 1,6-bisphosphate D-glyceraldehyde 3-phosphate-lyase   EC 4.1.2.13

   fructose 1,6-bisphosphate ↔ glyceraldehyde-3-phosphate + dihydroxyacetone
   phosphate

3. Fructose 1,6-bisphosphatase
   D-fructose 1,6-bisphosphate 1-phosphohydrolase   EC 3.1.3.11

   fructose 1,6-bisphosphate + H_2O → fructose 6-phosphate + P_i

4. Phosphoglucose isomerase
   D-glucose 6-phosphate ketolisomerase   EC 5.3.1.9

   glucose 6-phosphate ↔ fructose 6-phosphate

5. Phosphoglucomutase
   α-D-glucose-1,6-phosphomutase:α-D-glucose 1-phosphate   EC 5.4.2.2

   glucose 6-phosphate ↔ glucose 1-phosphate

6. UDP-glucose pyrophosphorylase
   UTP:α-D-glucose 1-phosphate uridylyltransferase   EC 2.7.7.9

   UTP + glucose 1-phosphate ↔ UDP-glucose + PP_i

7. Sucrose phosphate synthase
   UDP-D-glucose:D-fructose 6-phosphate 2-α-D-glycosyltransferase   EC 2.4.1.14

   UDP-glucose + fructose 6-phosphate → sucrose 6-phosphate + UDP

8. Sucrose phosphatase
   Sucrose 6-phosphate phosphohydrolase   EC 3.1.3.24

   Sucrose 6-phosphate + H_2O → sucrose + P_i

9. Pyrophosphate:fructose 6-phosphate transferase
   Pyrophosphate:D-fructose 6-phosphate 1-phosphotransferase   EC 2.7.1.90

   PP_i + fructose 6-phosphate ↔ P_i + fructose 1,6-bisphosphate

10. Fructose 6-phosphate,2-kinase
    ATP:D-fructose 6-phosphate 2-phosphotransferase   EC 2.7.1.105

    Fructose 6-phosphate + ATP → fructose 2,6-bisphosphate + ADP

11. Fructose 2,6-bisphosphatase
    D-fructose 2,6-bisphosphate 2-phosphohydrolase   EC 3.1.3.46

    Fructose 2,6-bisphosphate + H_2O → fructose 6-phosphate + P_i

need to retain a proportion of the triose phosphate to generate the carbon dioxide acceptor, ribulose 1,5-bisphosphate. An unregulated sucrose synthesis would deplete the chloroplast of intermediates of the Calvin cycle, so that photosynthesis would come to a halt. On the other hand, under optimum conditions for photosynthesis the demand for sucrose from the physiological sink could be exceeded and an unrestrained synthesis of sucrose would result in an accumulation of sucrose with undesirable consequences for the osmotic

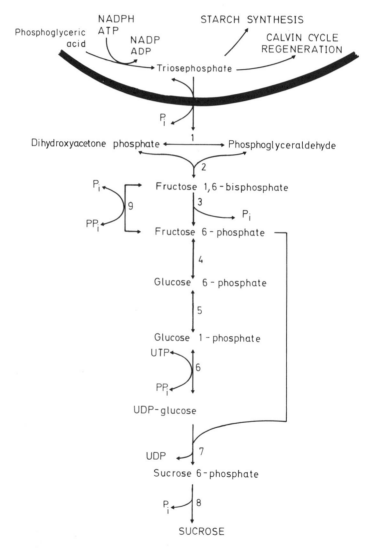

Fig. 2.4   Biosynthetic pathway for sucrose synthesis in photosynthetic cells. The numbers refer to the enzymes shown in Table 2.2.

relations of the photosynthetic tissue. Under these conditions there would obviously be a need to divert assimilate away from the synthesis of sucrose and towards the synthesis of starch in the chloroplast stroma.

When the standard free-energy changes of the reactions involved in the pathway from triose phosphate to sucrose are considered, it becomes apparent that there are effectively three irreversible reactions: those catalysed by the enzymes: fructose 1,6-bisphosphatase, sucrose phosphate synthase and sucrose phosphatase. The potential for regulation at these points in the pathway has been confirmed by much biochemical work which has shown that fructose 1,6-bisphosphatase and sucrose phosphate synthase are subject to a coordinated regulation which controls the rate of sucrose synthesis.

The fructose 1,6-bisphosphatase activity is controlled principally by the concentration of the regulatory metabolite fructose 2,6-bisphosphate (Figure 2.5).

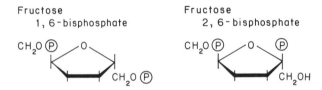

Fig. 2.5 The structures of fructose 1,6-bisphosphate and fructose 2,6-bisphosphate.

Concentrations of this potent inhibitor of the fructose 1,6-bisphosphatase change markedly in response to changes in the levels of sucrose, either when the sucrose is supplied exogenously or when sucrose is allowed to accumulate in photosynthesising leaf tissue. Thus as the photosynthetic rate is increased, either by making more light available or by raising the carbon dioxide concentration, the fructose 2,6-bisphosphate level is lowered, in this way deregulating the fructose 1,6-bisphosphatase and favouring sucrose synthesis. Conversely, if during photosynthesis export of sucrose from the leaf is prevented, the level of fructose 2,6-bisphosphate is raised so that assimilate is diverted away from sucrose synthesis and towards the accumulation of starch.

The concentration of fructose 2,6-bisphosphate is adjusted *in vivo* by the relative activities of two enzymes: fructose 6-phosphate, 2-kinase, which is responsible for the synthesis of fructose 2,6-bisphosphate from fructose 6-phosphate and ATP; and fructose 2,6-bisphosphatase, which is responsible for the removal of phosphate from carbon 2 of fructose 2,6-bisphosphate with the regeneration of fructose 6-phosphate (Figure 2.6).

The concentrations of fructose 2,6-bisphosphate in the cytosol are within the range 1 to 10 μM, while the concentrations of fructose 6-phosphate and fructose 1,6-bisphosphate are nearer 1 mM. Therefore the amounts of phosphorylated fructose diverted from the main pathway to furnish fructose 2,6-bisphosphate do not significantly deplete the main assimilate flow to sucrose.

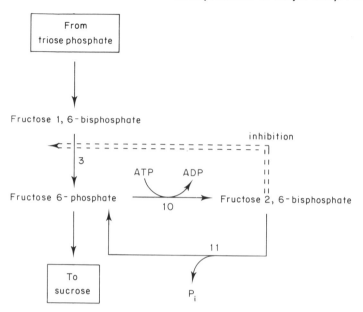

Fig. 2.6 The synthesis and breakdown of fructose 2,6-bisphosphate in the pathway of sucrose synthesis. The numbers refer to the enzymes shown in Table 2.2.

The concentration of fructose 2,6-bisphosphate changes during the course of the day. As photosynthesis starts up in the morning, there is an initial decrease in the concentration of fructose 2,6-bisphosphate. Then during the day its concentration slowly increases. These changes in the concentration of fructose 2,6-bisphosphate during the course of the day can easily be related to the ratio of activities of the kinase and phosphatase.

The enzymes which regulate the fructose 2,6-bisphosphate level have been shown to be sensitive *in vitro* to the interaction of four metabolites; fructose 6-phosphate, phosphate, phosphoglyceric acid, and dihydroxy-acetone phosphate. Fructose 6-phosphate and phosphate activate the fructose 6-phosphate, 2-kinase and inhibit the fructose 2,6-bisphosphatase. The fructose 6-phosphate, 2-kinase is inhibited by phosphoglyceric acid and dihydroxyacetone phosphate, but the inhibition is overcome by increased concentrations of fructose 6-phosphate. These effects are summarised in Table 2.3.

Fructose 2,6-bisphosphate has an additional controlling effect on the sucrose synthesis pathway via another enzyme: the pyrophosphate:fructose 6-phosphate transferase. This enyzme catalyses the reversible phosphorylation of fructose 6-phosphate by pyrophosphate to re-form fructose 1,6-bisphosphate. Its catalytic activity in the direction of fructose 1,6-bisphosphate formation is activated by nM concentrations of fructose

Sucrose

Table 2.3 Metabolite effectors of the two enzymes in spinach leaf responsible for controlling the fructose 2,6-bisphosphate concentration in spinach leaves (from Stitt *et al.*, 1987, with permission).

Enzyme	Activators	Inhibitors
Fructose 6-phosphate, 2-kinase	Fructose 6-phosphate Phosphate	Phosphoglyceric acid Dihydroxyacetone phosphate
Fructose 2,6-bisphosphatase		Fructose 6-phosphate Phosphate

2,6-bisphosphate, while its activity in the opposite direction, the formation of pyrophosphate, is activated by μM concentrations of fructose 2,6-bisphosphate. Thus low fructose 2,6-bisphosphate concentrations (which permit sucrose synthesis) favour the removal of the pyrophosphate formed by the sucrose phosphate synthase reaction.

Sucrose phosphate synthase is an equally important control point in the biosynthesis of sucrose. It is subject to two types of control: a fine control by which effector metabolites exert an immediate effect; and a coarse control in which slower changes in the activity of the enzyme are apparent when the enzyme is extracted.

Fine control takes the form of an activation by glucose 6-phosphate and an inhibition by phosphate, so that the enzyme activity depends on the glucose 6-phosphate:phosphate ratio. These regulatory effects are likely to be due to allosteric effects, since treatments with reagents that attack sulphydryl groups on proteins cause the enzyme to become independent of glucose 6-phosphate and phosphate without losing its catalytic activity. Sucrose phosphate synthase is also inhibited by sucrose, its sensitivity varying between different species.

The coarse control of sucrose phosphate synthase is not well characterised biochemically, but probably involves the reversible addition of a phosphate group to the enzyme protein. It is known that the activity of the extracted enzyme responds to both the input of assimilate into the sucrose synthesis pathway and the output of assimilate from the pathway. Thus activity is enhanced by illuminating the plant, or by partial defoliation which stimulates assimilate demand from the remaining leaves; and activity declines when the export of sucrose is blocked. In addition to these effects, the activity of the sucrose phosphate synthase changes according to a diurnal rhythm. This appears to be an endogenous rhythm controlled by an internal clock, and independent of environmental changes. At the onset of illumination the extractable activity increases sharply, to decline slowly during the period of illumination.

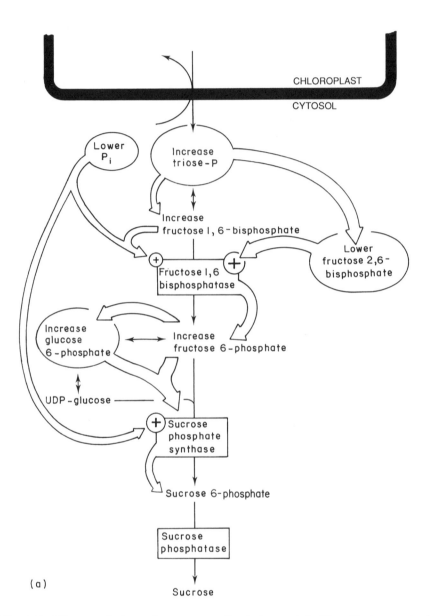

Fig. 2.7 The regulation of sucrose synthesis by metabolite effector control of fructose 1,6-bisphosphatase and sucrose phosphate synthase. (a) Feedforward control in response to an increased flow of assimilate from the chloro-

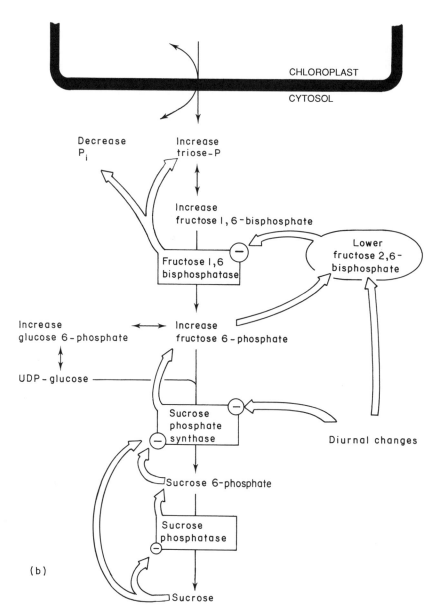

plast; (b) feedback control in response to an 'overproduction' of sucrose. (Modified from Stitt *et al.*, 1987. Reproduced by kind permission of the authors and of the publishers, Academic Press Inc.)

The sequence of regulatory events which would result in an increased flow of sucrose from an increased rate of photosynthesis are shown in Figure 2.7a. A greater flow of triose phosphate from the chloroplast withdraws phosphate from the cytosol, and reduces the concentration of fructose 2,6-bisphosphate, thus stimulating the fructose 1,6-bisphosphatase. This increases the concentrations of fructose 6-phosphate and glucose 6-phosphate which combine with the lowered phosphate concentration to stimulate the sucrose phosphate synthase. If sucrose begins to accumulate in the source leaf (Figure 2.7b) it would start to inhibit the sucrose phosphate synthase and sucrose phosphatase, and the effect would be transmitted back along the pathway, initially via a rising fructose 6-phosphate concentration. A similar effect would be transmitted if the sucrose phosphate synthase activity were declining due to the endogenous rhythm (Figure 2.7b). This would increase the level of fructose 2,6-bisphosphate, with an immediately negative effect on the fructose 1,6-bisphosphatase. The resulting build up of fructose 1,6-bisphosphate would be sensed as an accumulation of triose phosphate back in the chloroplast stroma where it would be channelled to starch synthesis.

When experiments have been carried out *in vitro* with metabolite concentrations similar to those found to occur in the cytosol of photosynthesising leaves, the activities of the key regulatory enzymes fructose 1,6-bisphosphatase and sucrose phosphate synthase have been seen to be sensitive to a concentration range of key effector metabolites, like dihydroxyacetone phosphate, not in a linear way; rather the enzyme is activated when the effector reaches a threshold concentration. The result of this threshold effect is that sucrose synthesis is active only when the chloroplast is supplying triose phosphate over and above the maintenance requirements of the cell. It is the surplus assimilate which is directed towards sucrose synthesis.

SUCROSE ACCUMULATION

The sucrose made in the leaves is transported to the storage tissues in the sugar cane and sugar beet by the phloem. When the sucrose is unloaded at the storage site there are two possible routes by which it could be transferred to the storage cells: symplastically, that is via the plasmodesmata that penetrate the cell walls of adjacent cells and interconnect cytosolic compartments; or apoplastically, that is via the physiological free space of the cell wall. Which of the two routes is actually taken? On the one hand, it is not always easy to observe the appropriate plasmodesmatal communication; on the other hand, the apoplastic route is more complicated, involving an outward transport across the plasma membrane of the phloem cells and an inward transport across the plasma membrane of the storage cells. A general conclusion seems to be that the symplastic route is taken by sucrose unloaded into actively growing organs such as young leaves and roots, while sucrose is unloaded into the apoplast for reabsorption by mature, storage organs such as cane stems and beet roots (Lucas & Madore, 1988).

In both sugar cane and sugar beet sucrose is accumulated in a specialised tissue, the storage parenchyma, made up of large, thin-walled cells. The final sucrose concentration reached in both cases is about 20 per cent of the fresh weight. In sugar cane some of this sucrose may be in the cell wall space, but most of the stored sugar must be located in the cell vacuoles, which provide by far the largest available compartment. The final concentration within the vacuoles can be as high as 900 mM.

Sucrose is the only one of the major crop products described here that is not in a polymeric form (stretching the definition of a polymer to include the fatty acids of oils). Thus while in the biosynthesis of the other crop products one of the final steps takes the form of a polymerisation reaction, in the case of sucrose accumulation the final step is the concentration of sucrose into the storage compartment. This is essentially a membrane transport process. So while the deposition of the other crop products could be described in largely biochemical terms, the accumulation of sucrose is more of a biophysical process.

In sugar cane the storage parenchyma in the stems is interspersed with vascular tissue, and surrounded by a hard, wax-covered rind. When internodes are still growing, there is an acid invertase activity associated with the sap squeezed from the cane. Most of this sap is made up of the contents of the vacuole, which is at about pH 5.5, the pH optimum of the invertase. Thus the soluble acid invertase is more likely to be located in the vacuole than in the cytosol, which is expected to be at about pH 7.5. The probable function of this vacuolar invertase in the immature internodes is to catalyse the initial step in the mobilisation of sucrose to support growth. When the internode stops growing, the high-yielding varieties of sugar cane lose acid invertase activity; varieties that retain an invertase activity in the mature internodes being unable to store high levels of sucrose.

Towards the end of the growing period there is an increase in the sucrose content of the canes associated with the cooler and drier conditions which reduce growth but allow photosynthesis to continue. During the final ripening the canes lose water, and this partial desiccation further increases the sucrose content measured on a fresh weight basis.

While in the sugar cane stem the units of construction are the internodes, in the roots of sugar beet the equivalent component units would be the concentric rings of tissue which are apparent when the swollen root is cut transversely. These rings are due to an alternation of layers of large-celled storage parenchyma cells, with layers of tissue containing small-celled parenchyma and vascular bundles. The rings are derived from secondary meristems which are laid down in the primary roots as a series of concentric rings outside the pericycle. Continuous cell multiplication and expansion from the meristematic rings over the growing season enlarges the storage root. The rings develop simultaneously rather than sequentially, but there are rings which are

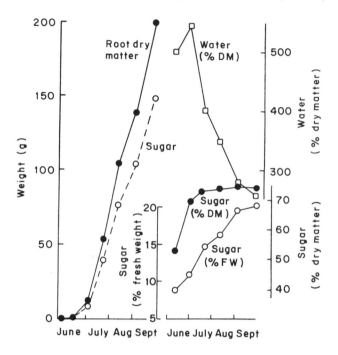

Fig. 2.8   Changes in the sugar content in relation to the growth of the storage roots of sugar beet. (From Milford, 1973. Reproduced by kind permission of the author and of the Association of Applied Biologists.)

initiated later in the season, and these remain immature at harvest (Milford, 1973).

During the growing season the sucrose concentration of the beet root increases gradually on a fresh weight basis, but as a percentage of the dry matter, the sucrose content attains its final value of around 70 per cent of the dry weight early in the season (Figure 2.8).

As in sugar cane, an invertase activity is apparent in the very earliest developmental stages of beet root, but it is replaced later by sucrose synthase, which remains throughout the period of sucrose accumulation. Presumably the invertase and sucrose synthase function, as in cane, is the mobilisation of incoming sucrose for biosynthesis. The problem arises, however, of reconciling the continued presence of a sucrose synthase with the accumulation of stored sucrose. How does the incoming sucrose avoid being metabolised? The storage tissue of the beet roots is not a uniform tissue, and one possibility is that the sucrose synthase is confined to cells which are at the very earliest stages of growth and consequently are constructing cell walls and other cell components rather than accumulating sucrose. These cells would be mobilising sucrose rather than storing it.

The earliest experiments on the storage of sucrose were carried out with thin discs (0.5 or 1 mm in thickness) cut from the storage parenchyma of sugar cane and beet. Uptake of sugars into the discs was followed when the discs were fed ^{14}C-labelled sugars, and the pattern of labelling in the stored sugars was analysed. In addition the enzymes present in the discs were assayed. The work with sugar cane was started in the 1960s by investigators in Australia (Glasziou & Gayler, 1972). Their findings can be summarised as follows (Figure 2.9).

Sucrose is hydrolysed in the free space by an acid hydrolase located at the cell wall. Glucose and fructose are taken up independently by separate carriers in the plasma membrane, but sucrose itself cannot be taken up. In the cytosolic compartment sucrose phosphate is synthesised by the activity of sucrose phosphate synthase. Sucrose is stored in the vacuole. It was suggested that the movement of sucrose into the vacuole might be associated with the activity of the sucrose phosphatase located at the tonoplast. This was a speculative proposal, since cell fractionation had not yet established the subcellular localisation of the sucrose phosphatase.

The hydrolysis of sucrose in the free space outside the plasma membrane was attributed to an acid invertase which was observed in washed residues of cell extracts. When its activity in the discs was inhibited by addition of Tris buffer to the suspending medium, uptake of label from sucrose was inhibited.

The requirement for sucrose hydrolysis was also supported by experiments in which sucrose radioactively labelled in the fructosyl units was supplied to the discs, and the stored sucrose was analysed for the distribution of label in the glucosyl and fructosyl units (Figure 2.10). This experiment allows us to decide whether the sucrose supplied is hydrolysed and resynthesised, or if it is taken up without hydrolysis. If the sucrose is hydrolysed, the released hexoses can be phosphorylated by hexokinases, and the resulting glucose 6-phosphate and fructose 6-phosphate can be converted one to another by the action of phosphoglucomutase. Subsequent resynthesis of sucrose from this pool of hexose phosphates would yield sucrose in which the label is equally distributed between the glucosyl and fructosyl units. On the other hand, in the absence of hydrolysis and resynthesis no equilibration between the glucosyl and fructosyl units could occur, and consequently the stored sucrose would retain the asymmetric distribution of label in the sucrose supplied.

The results obtained with cane discs show that there was a substantial redistribution of label from the fructosyl to the glucosyl unit of the stored sucrose. By contrast the results obtained when a similar experiment was carried out with discs cut from sugar beet roots showed no significant redistribution of label (Table 2.4). In the case of beet, labelled glucose or fructose was also supplied alone, in order to ensure that isomerisation could have occurred if the asymmetrically labelled sucrose supplied had been broken down. The ^{14}C-glucose/^{14}C-fructose ratio near to unity when the labelled hexoses were supplied (Table 2.4) indicates that the recovered sucrose was labelled to a similar extent in the two halves and thus that isomerisation was possible.

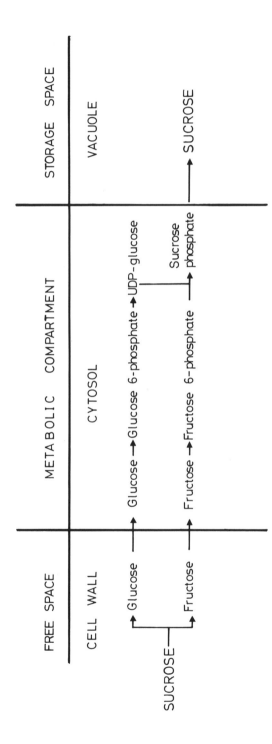

Fig. 2.9 The pathway of sucrose uptake into storage parenchyma cells of sugar cane, based on experiments using thin discs (based on the work of Glasziou & Gayler, 1972).

A. Sucrose hydrolysed during uptake

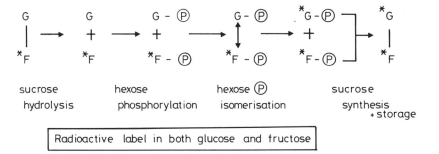

sucrose   hexose   hexose Ⓟ   sucrose
hydrolysis phosphorylation isomerisation   synthesis
                                            + storage

Radioactive label in both glucose and fructose

B Sucrose stored without hydrolysis

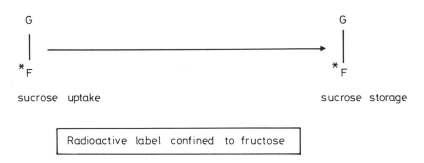

sucrose uptake   sucrose storage

Radioactive label confined to fructose

Fig. 2.10 The effect of hydrolysis on the intramolecular distribution of labelling in stored sucrose derived from asymmetrically labelled sucrose.

Table 2.4 The ratio of radioactive label in the glucose and fructose moieties of sucrose after accumulation of ^{14}C-labelled sugars by tissue discs cut from sugar beet roots and sugar cane stems (data from Giaquinta, 1977).

^{14}C-labelled sugar supplied	^{14}C-glucose/^{14}C-fructose ratio	
	Beet	Cane
Sucrose [^{14}C-fructose]	0.007	0.2–0.25
Sucrose [^{14}C-fructose + ^{14}C-glucose]	1	
Glucose	1.2	
Fructose	0.7	

Support for the role of the cell wall invertase in the uptake of sucrose in sugar cane has come from experiments using cells grown in suspension culture (Komor *et al.*, 1981). These cells were derived from sugar cane storage parenchyma, and are reported to retain the ability of the original tissue to accumulate sucrose to a concentration of 20 per cent of the fresh weight and 60 per cent of the dry weight. Like the discs, the cells take up glucose and fructose, but not sucrose. Radioactive label was accumulated from sucrose only after the sucrose had been hydrolysed by an acid invertase in the cell wall. In agreement with the proposed role of the cell wall invertase, uptake of radioactivity from sucrose into protoplasts, prepared by digesting the cell wall, was very much reduced compared to uptake into the complete cells.

The experimental findings with the discs and suspension culture cells have led to the general acceptance of the view that sugar beet roots accumulate sucrose without hydrolysis, while in sugar cane a preliminary hydrolysis of sucrose to glucose and fructose for sucrose accumulation is necessary. In sugar beet the wall invertase activity apparent in the younger, immature roots disappears as the root matures. Moreover a synthetic analogue of sucrose, 1-fluorodeoxysucrose, is accumulated in the vacuoles of sugar beet roots as readily as sucrose even when it is supplied to the leaves. Since this occurs despite the fact that the sucrose analogue is a poor substrate for invertase (Lemoine *et al.*, 1988), there can be little doubt that sucrose is stored in the roots of sugar beet without hydrolysis.

However, doubts can be entertained over the requirement for hydrolysis in the storage of sucrose by sugar cane. First, uptake of asymmetrically labelled sucrose into discs cut from mature (non-elongating) cane internodes can be observed to occur with less than 15 per cent of the label redistributed within the sucrose molecule (Lingle, 1989). Thus it is possible for sucrose to be stored without hydrolysis. Second, in some tissues it is known that acid invertase activity can be elicited by mechanical injury, and thus its activity in discs may be more a consequence of the preparation of the discs, rather than being a feature of the tissue *in vivo*. Similarly one can question the relevance of the wall invertase observed in the suspension cultured cells to the organised storage parenchyma tissue of the mature internode. Third, in beet it has been found that the uptake of sucrose into discs is sensitive to tissue turgor (Wyse *et al.*, 1986), so that simply rinsing the sugar cane discs in media lacking an appropriate osmotic support could have inactivated a sucrose uptake system that may be present in cane. Finally, the physiologically defined free space becomes a relatively large proportion of the volume of the discs compared to the proportion it occupies in the intact tissue, simply because of the damage incurred by cells at the disc surface during slicing. Taken together, these cautionary points suggest that studies of sucrose uptake into storage parenchyma of cane may have over-emphasised the cell wall and its invertase activity, and under-estimated the ability to take up sucrose unhydrolysed. It is therefore uncertain whether hydrolysis of sucrose is required for its accumulation in stems of sugar cane.

The accumulation of sucrose in the storage vacuoles can be viewed as the result of the operation of two sets of pumps working in series: the first set at the plasma membrane, the second at the tonoplast. Characterisation of sucrose transport at these two membranes therefore seemed to investigators to be of paramount importance in understanding sucrose storage at a biochemical level.

In sugar cane, transport of sugars has been studied in cells grown in suspension culture, and in vacuoles isolated from protoplasts prepared from these cells. As described earlier, uptake of sucrose is not observed with these cells, despite their ability to accumulate sucrose to levels as high as those of the storage parenchyma from which they are derived. Glucose and fructose are, however, taken up (Komor *et al.*, 1981). The concentrative ability of this transport could be estimated when a non-metabolisable analogue of glucose, 3-*O* methylglucose, was presented to the cells. This compound was accumulated to an intracellular concentration 15 times that in the medium. By using a pH meter in the medium in which the cells were suspended, it was found that uptake of glucose (and 3-*O* methylglucose) was accompanied by the uptake of protons into the cells. It was proposed that hexose uptake was coupled to proton transport with a stoichiometry of one H^+ per sugar molecule. This type of transport is an example of a proton symport. The plasma membrane of plant cells generally possesses a H^+-translating ATPase, the activity of which leads to protons being translocated outwards across the plasma membrane, thus generating a pH gradient (acid outside) and a membrane potential (positive outside). Together the pH gradient and membrane potential constitute a proton-motive force which would drive sugar uptake by the proposed proton symport. It is probable that all plant cells have mechanisms similar to this for the uptake of sugars. The relevance of hexose transport to the accumulation of sucrose in sugar cane hinges on whether hydrolysis of sucrose to its constituent is required for sucrose accumulation, as discussed above.

The vacuoles isolated from the suspension cultured cells of sugar cane took up glucose (and 3 *O*-methylglucose) and sucrose (Thom *et al.*, 1982), but the rates of uptake with sucrose were an order of magnitude lower than the rates with glucose. Like the plasma membrane, the tonoplast possessed a H^+-translocating ATPase—in this case driving protons into the vacuole—and glucose uptake was accompanied by a counter movement of one H^+ per glucose molecule (Thom & Komor, 1984). So, as with the plasma membrane, glucose uptake across the tonoplast appeared to be driven by an ATPase generated proton-motive force. The rates of sucrose uptake were inadequate to account for the rates at which sucrose was accumulated in vacuoles of protoplasts supplied with ^{14}C-labelled glucose, and it was concluded that 'either isolation or the incubation procedure appears to be unfavourable for sucrose transport in the isolated vacuoles' (Thom *et al.*, 1982).

Discs prepared from sugar beet roots can be shown to have a sucrose carrier operating at the plasma membrane, if the discs are offered an appropriate osmotic support in their suspending medium; the inert sugar alcohol, mannitol at 400 mM works well (Wyse *et al.*, 1986). The apparent K_m of the carrier for sucrose is about 20 mM, and uptake is inhibited by the addition of low concentrations (2.5 μM) of carbonylcyanide-*m*-chlorophenylhydrazone (CCCP). This compound increases the permeability of membranes to protons, and therefore it is likely that sucrose uptake across the plasma membrane of parenchyma cells from sugar beet roots is driven by an ATPase-generated proton-motive force.

Vacuoles isolated from beet roots have been shown to possess a proton-translocating ATPase (and pyrophosphatase), but an ATP-dependent accumulation of sucrose has not proved to be readily demonstrable. However, valuable information has come from experiments which have used the small

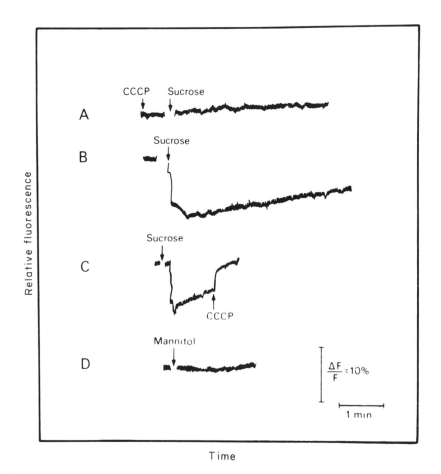

Time

Fig. 2.11 Membrane potential changes induced by the addition of sucrose to tonoplast vesicles prepared from sugar beet roots. A fluorescent dye added to the suspension of tonoplast vesicles responds to changes in the membrane potential by changes in the fluorescence emitted (F). (From Briskin *et al.*, 1985. Reproduced by kind permission of the authors and of the American Society of Plant Physiologists.)

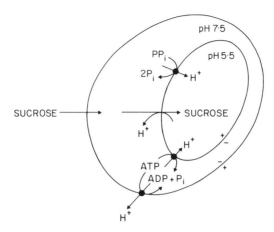

Fig. 2.12 Transport across the plasma membrane and tonoplast in the accumulation of sucrose by sugar beet storage parenchyma cells.

vesicles which reform from fragments of tonoplast produced when the root is homogenised (Briskin *et al.*, 1985). Uptake of sucrose into these tonoplast vesicles is largely dependent on ATP, and appears to occur via a saturable carrier (K_m 12 mM). When sucrose is added to the vesicles a transient change in the membrane potential (interior more negative) can be detected (Figure 2.11). All these effects are sensitive to CCCP. Thus in roots of sugar beet there is direct evidence that ATP can drive sucrose uptake into the vacuoles via a H^+/sucrose antiport at the tonoplast (Figure 2.12).

Much excitement was created when Thom and Maretzski (1985) found what appeared to be evidence for an enzyme complex at the tonoplast of sugar cane which synthesised sucrose from the glucosyl group of UDP-glucose. They proposed that this group translocator, as the complex was termed, catalysed a vectorial synthesis of sucrose, taking its substrate from the cytosol and depositing the sucrose product within the vacuole. However, it was subsequently revealed (Preisser & Komor, 1988) that the disaccharide formed in their experiments had been misidentified: it was laminaribiose, a β-1,3-glucan; no sucrose had been made. Most of the glucose incorporated from the UDP-glucose supplied as substrate to the vacuoles was incorporated into callose by the activity of a callose synthase (glucan synthase II, see Chapter 5), which is located at the plasma membrane, and therefore will contaminate preparations of vacuoles such as those used by Thom & Maretzki (1985) that contain significant amounts of plasma membrane.

BIOTECHNOLOGICAL DEVELOPMENT

Unlike many of the other crop products, sucrose cannot be 'improved' chemically. Where there is room for progress is in the yield that can be obtained and in the removal from the sucrose-producing plants of secondary compounds which interfere with the recovery of sucrose.

In beet an increased yield is likely to come from two approaches. First, it has long been recognised that the potential yield of sugar beet is limited by the lateness with which the foliage develops from a spring sowing. Complete cover of the soil does not normally occur until after late spring, so that full advantage cannot be taken of the solar radiation available in late spring. One solution to this problem would be to develop cold-hardy, bolting-resistant varieties suitable for sowing in autumn.

Second, knowledge of the biochemical basis of the regulation of sucrose biosynthesis is now providing the kind of information which should soon enable us to identify precisely the enzymes which should be modified in order to maximise sucrose production. Thus for example, the sensitivity of the cytosolic fructose 1,6-bisphosphatase to the regulatory fructose 2,6-bisphosphate varies between different crop plants such as wheat and maize. A lowered sensitivity of the leaf fructose 1,6-bisphosphatase to the regulatory metabolite, fructose 2,6-bisphosphate would allow a greater flow of assimilate towards sucrose, which could be advantageous for a plant like sugar beet. In addition genetic interference with the covalent modifications of the leaf sucrose phosphate synthase (Huber & Huber, 1990) offer further scope for the deregulation of sucrose synthesis.

Past selection in the development of sugar cane and sugar beet has given us varieties which have a high potential for sugar storage. The invertase activity of these varieties during sugar storage is low, its development being delayed until regrowth creates a demand within the plant for the energy and carbon represented by the accumulated sugar. The low reducing sugar levels are an advantage, because during processing sucrose crystallisation is hindered by reducing sugars. However, there are other constituents that interfere with extraction, particularly nitrogen-containing compounds accumulated by sugar cane and sugar beet. Possibly these could be removed genetically with benefits to the efficiency of sucrose extraction. The betaine which is accumulated by sugar beet is a notable example of such a nitrogen-containing compound.

FURTHER READING

Avigard, G. (1982). Sucrose and other disaccharides, in *Encyclopaedia of Plant Physiology, Vol 13A Plant Carbohydrates I Intracellular Carbohydrates*, Eds Loewus, F.A. & Tanner, W., Berlin, Springer Verlag, pp. 217–347.

Blackburn, F. (1984). *Sugar-cane*. London, Longman.

Glasziou, K.T. & Gayler, K.R. (1972). Storage of sugars in stalks of sugar cane, *Botan. Rev.* **38**, 471–490.

Hawker, J.S. (1985). Sucrose, in *Biochemistry of Storage Carbohydrates in Green Plants*, Eds Dey, P.M. & Dixon, R.A., London, Academic Press, pp. 1–51.

Huber, S.C. & Huber, J.L.A. (1990). Regulation of spinach leaf sucrose phosphate synthase by a multisite phosphorylation, *Curr. Top. Plant Biochem. Physiol.* **9**, 329–343.

Stitt, M., Huber, S. & Kerr, P. (1987). Control of photosynthetic sucrose formation, in *The Biochemistry of Plants. Volume 10 Photosynthesis*, Eds Hatch, M.D. & Boardman, N.K., London, Academic Press, pp. 327–409.

Wayman, M. & Parekh, S.R. (1990). *Biotechnology of Biomass Conversion.* Milton Keynes, Open University Press.

Yudkin, J., Edelman, J. & Hough, L. (1971). *Sugar.* London, Butterworths.

ADDITIONAL REFERENCES

Anon. (1984). *Biomass for Energy.* Paris, OECD.

Briskin, D.P., Thornley, W.R. & Wyse, R. (1985). Membrane transport in isolated vesicles from sugarbeet taproot. II. Evidence for a sucrose/H^+-antiport, *Plant Physiol.* **78**, 871–875.

Fondy, B.R. & Geiger, D.R. (1982). Diurnal pattern of translocation and carbohydrate metabolism in source leaves of *Beta vulgaris* L, *Plant Physiol.* **70**, 671–676.

Giaquinta, R. (1977). Sucrose hydrolysis in relation to phloem translocation in *Beta vulgaris, Plant Physiol.* **60**, 339–343.

Komor, E., Thom, M. & Maretzki, A. (1981). The mechanism of sugar uptake by sugarcane suspension cells, *Planta* **153**, 181–192.

Lemoine, R., Daie, J. & Wyse, R. (1988). Evidence for the presence of a sucrose carrier in immature sugar beet tap roots, *Plant Physiol.* **86**, 575–580.

Lingle, S.E. (1989). Evidence for the uptake of sucrose intact into sugarcane internodes, *Plant Physiol.* **90**, 6–8.

Lucas, W.J. & Madore, M.A. (1988). Recent advances in sugar transport, in *The Biochemistry of Plants, Vol 14 Carbohydrates*, Ed Preiss, J., New York, Academic Press, pp. 35–84.

Milford, G.F.J. (1973). The growth and development of the storage root of sugar beet, *Ann. Appl. Biol.* **75**, 427–438.

Preisser, J. & Komor, E. (1988). Analysis of the reaction products from incubation of sugarcane vacuoles with uridine diphosphate-glucose: no evidence for the group translocator, *Plant Physiol.* **88**, 259–265.

Thom, M. & Komor, E. (1984). H^+-sugar antiport as the mechanism of sugar uptake by sugarcane vacuoles, *FEBS Lett.* **173**, 1–4.

Thom, M. & Komor, E. (1985). Electrogenic proton translocation by the ATPase of sugarcane vacuoles, *Plant Physiol.* **77**, 329–334.

Thom, M., Komor, E. & Maretzki, A. (1982). Vacuoles from sugarcane suspension cultures II. Characterization of sugar uptake, *Plant Physiol.* **69**, 1320–1325.

Thom, M. & Maretzki, A. (1985). Group translocation as a mechanism for sucrose transport into vacuoles from sugarcane cells, *Proc. Natl. Acad. Sci. USA* **82**, 4697–4701.

Wyse, R.E., Zamski, E. & Tomos, A.D. (1986). Turgor regulation of sucrose transport in sugar beet taproot tissue, *Plant Physiol.* **81**, 478–481.

Chapter 3

Starch

INTRODUCTION

Starch is the principal reserve carbohydrate of plants. It is also one of the most characteristic plant products, with virtually every cell in every higher plant able to produce it. Within the plant cells, starch is always found in the form of granules (also called starch grains) contained within plastids.

Starch is a polysaccharide constructed from glucose as the basic building block. It is a mixture of two α-glucan polymers: amylose, in which the glucosyl units are joined by α-1,4 linkages to form essentially unbranched chains up to several thousand units long; and amylopectin, in which shorter α-1,4 chains are connected by α-1,6 linkages to form larger, highly branched molecules containing some 50 000 glucosyl units (Figure 3.1).

Starch functions in plants both as a short-term and as a long-term store of carbon and energy. In the chloroplasts of photosynthetic tissues it undergoes a diurnal rhythm of accumulation and breakdown. In non-photosynthetic tissues it is deposited for longer periods: in underground storage organs the starch is re-utilised in the following season; in dry seeds it can remain available for years. In these longer-term stores starch is accumulated to form a high proportion of the dry matter, and it is in this form that starch becomes important as a component of the harvested crop.

From both the commercial and nutritional points of view, starch can be considered to be the single most valuable crop product. It forms about 70 per cent of the dry weight of the major cereal grains (Figure 3.2). In the principal tuber crops the starch content, as a percentage of the dry weight, varies from about 70 per cent in potato to about 90 per cent in cassava and sweet potato (Figure 3.2).

Starch is the major provider of calories in the diet in both temperate and

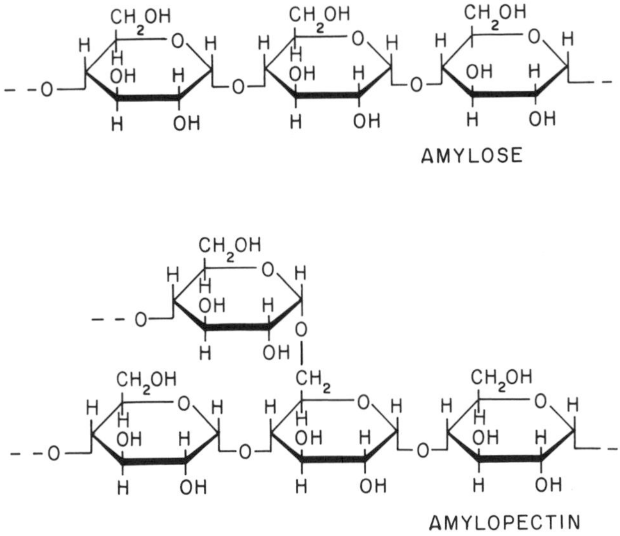

Fig. 3.1   The polymeric components of starch.

tropical regions. The annual production of cereals has been estimated to be some 1800 million tonnes of which starch makes up about 1000 million tonnes. Not only is starch important as a component of our principal food crops, large amounts are also extracted for industrial applications. Of the 4 million tonnes of starch produced each year in Western Europe (800 000 tonnes from potatoes), one-third is utilised by non-food industries, often after it has been improved by chemical or physical modification (Figure 3.3).

The starch granule represents a concentrated package of polymeric carbohydrate. It is insoluble in cold water and relatively dense, and it is for these reasons that it is easily purified: simply by washing it free from the soluble contaminants. Technically, starch has a number of advantages over the other polymeric carbohydrate produced by plants in a comparable quantity, cellulose. Starch, unlike cellulose, can be dispersed in cold water, which allows the physical structure of the starch granule to be easily modified. The glucosyl linkages in starch render it much more reactive than cellulose. This means that starch can more easily be modified chemically, both by derivatisation and by hydrolysis to the constituent oligosaccharides and glucose.

The food industry finds a multitude of uses for starch as a thickener, filler and binder. In the USA large quantities of maize starch are hydrolysed to glucose, which is then isomerised to fructose to form high-fructose corn syrup (HFCS). This product finds an ever-increasing number of applications in the manufacture of soft drinks and confectionery of all kinds (see Chapter 4). Demand for starch in both the native and modified form for non-food industries depends on the competitiveness of starch against synthetic alterna-

Biosynthesis of Major Crop Products

tives. However, looking to the future, a potentially important user of starch is predicted to be the synthetic polymer industry.

Addition of starch as an inert filler in polyvinylchloride and polyethylene plastics not only reduces their cost, but also promises to increase their biodegradability. The incorporation of 7–10 per cent starch into low-density polyethylene renders biodegradable even the ubiquitous polyethylene bag. An application for starch-polyvinylchloride has been in the manufacture of a water-soluble laundry bag for use in hospitals. The bag, containing the contaminated clothing, can be placed directly in the washing machine, where the bag dissolves. The advantage of using a water-soluble bag in this particular situation is that it eliminates handling and thus reduces risk of contamina-

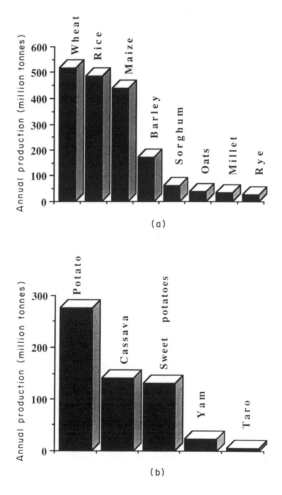

Fig. 3.2 Annual world production of the major starchy crops (average data for 1987–89; from FAO Production Yearbook, 1989).

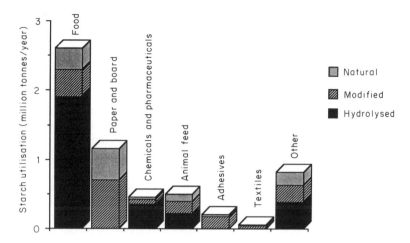

Fig. 3.3 The principal industrial uses of starch in the EEC in 1983 (data from Commission of the European Communities, 1984).

tion. In order to enhance the solubility of the starch for this application it is slightly modified chemically.

As well as acting as a filler, starch can also be used as an active component in a copolymer with a conventional synthetic polymer. Sometimes the copolymer is an improvement over the purely synthetic polymer. For example, the incorporation of a physically modified starch into polyurethane increases the effectiveness of the rigid polyurethane foam as an insulator. In this case the burning behaviour of the foam remains unchanged until the starch is chemically modified, when a significant reduction in flammability is obtained as an additional benefit.

Yet another approach to the incorporation of starch into synthetic polymers is through graft polymerisation. In this process the starch is allowed to react with the polymerisable vinyl or acrylic monomers so that the resulting polymer has a carbohydrate backbone with synthetic side chains. A wide variety of products with useful features can be made in this way, their properties depending on the nature of the synthetic polymer. One such graft polymer, already marketed in the USA, is made by graft polymerising acrylonitrile onto gelatinised starch. The dry polymer has a remarkable capacity to absorb water without dissolving, and has been given the apt name of Super Slurper. It finds a use in the manufacture of articles such as disposable nappies, where its ability to retain absorbed fluid under pressure is especially valuable.

Starch synthesis involves the construction of a predetermined structure, the starch granule, from two major components which amalgamate in a fixed ratio. Therefore before examining the enzymology of starch synthesis, we need to understand something of the structure and chemical composition of the starch granule itself.

THE STARCH GRANULE

Chemical composition

It has been recognised since the 1940s that starch is composed of two molecular species: the essentially linear amylose and its branched counterpart, amylopectin. They can be separated by leaching the amylose from starch granules with water, or by dissolving the starch granule in water and removing the amylose as an insoluble complex with a polar organic solvent such as *n*-propanol. The main properties of amylose and amylopectin are compared in Table 3.1.

Both amylose and amylopectin are soluble in water, but they differ in their behaviour towards aqueous solution. The relatively smaller amylose polymers are soluble in warm water, and will crystallise from solution if the temperature is lowered. However, if the concentration is too high or the temperature too low, the amylose precipitates from solution by a process called retrogradation. On the other hand, with amylopectin, once the molecules have been dispersed, solutions are stable.

Amylose is usually described as consisting of linear molecules, but it is now recognised that a limited amount of branching does occur. When purified amylose is subjected to hydrolysis by the enzyme α-amylase, complete hydrolysis does not occur, unless the enzyme pullulanase is also added. The α-amylase specifically attacks the α-1,4 linkages, while the pullulanase is able to hydrolyse the α-1,6 linkages. This requirement indicates the presence of some branching. It has been calculated that there are only 2–8 branch points per molecule, which is equivalent to about 1 branch point per 1000 glucose units. The branched nature of amylopectin is revealed by the residual dextrins of high M_r which remain when the α-1,4 links of the linear branches have been hydrolysed after attack by α-amylase. Branches occur about every 20 glucose units (5 per cent of the links). The chains that make up the amylopectin molecule vary in length from 12 to 60 glucoside units, with an average chain length of around 20 units.

Table 3.1 Properties of amylose and amylopectin.

	Amylose	Amylopectin
Structure	Essentially linear	Branched
Average chain length (glucosyl units)	ca.10^3	ca.20
M_r	$0.2–1 \times 10^6$	$10–500 \times 10^6$
Degree of polymerisation (glucosyl units)	$1.5–6 \times 10^3$	$0.3–3 \times 10^6$
Aqueous solution	Unstable	Stable
Iodine colour	Deep blue	Purple

The pattern of branching in the amylopectin molecule is believed to be as described in the cluster model of amylopectin structure (Figure 3.4), which takes its name from the clusters of interchain linkages which occur in specific regions of the macromolecule. In these regions, where the branches are concentrated, the chains are likely to be in an amorphous/gel structure, while in the regions where the chains are predominantly linear, with fewer branch points, there is probably a crystalline structure. Three catagories of chains have been distinguished within the amylopectin structure (Figure 3.4):

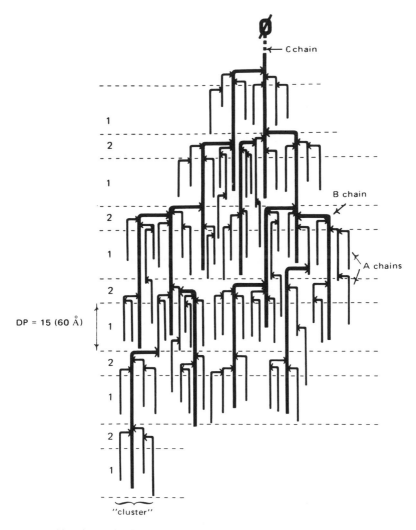

Fig. 3.4 The branched structure proposed for potato amylopectin. 1 = compact, crystalline region; 2 = less compact, amorphous region (from Robin *et al.*, 1974, reproduced with kind permission of The American Association of Cereal Chemists).

A-chains, short amylose chains substituted only at the reducing end where they are joined to the rest of the molecule by a single 1,6 bond; B-chains, substituted at the reducing end and at one or more C-6-OH groups by A-chains or by other B-chains; a single C-chain substituted at one or more C-6-OH groups, but unsubstituted at the reducing end. Each amylopectin molecule contains one C-chain, and thus one reducing group.

X-ray crystallography reveals that the essentially linear α-1,4-glucan chains of amylose and the branches of the amylopectin are helical with six glucosyl units per turn, and a diameter of 1.3 nm. It is the helical configuration of amylose that results in the deep blue (or blue-black) colour obtained when starch is exposed to iodine/iodide solution. The iodine molecule (actually as I_5^-, the polyiodide species) becomes trapped as a chain in the interior of the helix forming an inclusion complex. Purified amylopectin does not give the deep blue colour with iodine, presumably because the linear segments of the amylopectin molecule are too short for stable complexes of this type to form.

The proportions of amylose and amylopectin in starch are genetically determined, so that particular species or cultivars, will produce a starch with a particular ratio of amylose to amylopectin. In the major cereals—maize, wheat, barley, oats and rye—amylose constitutes 27–29 per cent of the starch. In rice the starch contains a higher proportion of amylose, which can reach 37 per cent, while the starch of potato contains a lower proportion of amylose at some 23 per cent. Mutant varieties of rice, maize, sorghum and pea are available in which this proportion can be altered drastically. Thus in maize the amylose content can be raised to 50 per cent in the mutant variety *amylose extender*, and lowered to zero in the mutant variety *waxy*. In varieties of wrinkle-seeded pea amylose is the major component and accounts for 66 per cent.

The starch obtained from cereal grains contains nearly 1 per cent dry weight as lipid. This lipid appears to be trapped within the helices of amylose, occupying the 'iodine site'. An association of the lipid with amylose explains why *waxy* varieties (zero amylose) have a much lower lipid content, and high amylose varieties have a higher lipid content. However, the absence of lipid from the starch of potatoes, cassava and leguminous seeds is not so easily explained. Another minor component of starch is phosphorus, which has been attributed to the phosphorylation of 1 in 300 glucose molecules. The significance of these minor components for the structure and synthesis of starch is not yet apparent.

Granule structure

The size and shape of starch granules are very variable, and depend on the species of plant that produced them. They can vary from 1–100 μm diameter; they can be spherical, oval lenticular or just irregular; they can be simple or complex aggregates. In fact the appearance of the granules under the light microscope is so characteristic it can be used to identify their botanical source. For example, potato starch granules have an average diameter of

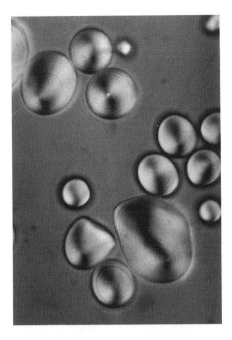

Fig. 3.5   Starch grains from potato recorded under interference contrast conditions (× 525). Courtesy of Dr M.J. Gidley, Unilever Research.

about 40 μm, are oval, and show concentric layers arranged around an eccentric spot, the hilum (Figure 3.5). In barley the starch granules are of two types: large lenticular granules of 15–35 μm diameter, and small spherical granules less than 10 μm in diameter. By weight about 90 per cent of barley starch is made up of the large granules, and by number about 90 per cent is made up by the small granules.

The internal organisation of the starch grain is not well understood. However, one principle is becoming widely recognised as an important feature of the architecture of the starch granule: its ordered, semi-crystalline three-dimensinal structure is given by the amylopectin component, rather than the amylose. There are three lines of evidence to support this principle:

1. In maize, the starch of the *waxy* mutant variety has the same crystallinity as the normal starch, yet has virtually no amylose. Conversely, the amylomaize starch, with a lowered amylopectin content, has a less ordered structure.
2. When starch granules, such as those of normal maize, are allowed to swell in hot water (gelatinisation) a large proportion of the amylose becomes solubilised and leached out, while the amylopectin remains—and the crystallinity is also retained.

3. When amylose is allowed to precipitate from cold water solution (retrogradation) subsequent solution in boiling water becomes very difficult. However, potato starch subjected to the same treatment shows no reluctance to return to solution in hot water. This difference can be accounted for by a different structuring of the amylose in the potato, presumably by its amylopectin partner.

It is paradoxical that the amylopectin component appears to confer the ordered crystallinity to the starch grain, since purified amylopectin is an amorphous non-crystalline polymer, while the virtually straight-chain amylose polymers can be readily crystallised from water.

When stained sections of starch granules are examined with the electron microscope radially arranged fibrils are seen. A general radial organisation of the starch granule is also seen when granules have been examined under the polarising microscope. It is likely that these fibrils represent radially extended starch molecules which appear to constitute the principal organisational unit of the starch granule. However, even the largest amylopectin molecules are not long enough to extend over the entire radius of a starch grain. Internally the starch granule shows a concentric arrangement of rings. These so-called growth rings can be seen by transmission and scanning electron-microscopy in granules which have been eroded by treatments with chemicals or enzymes. The rings represent layers of alternating high and low refractive index,

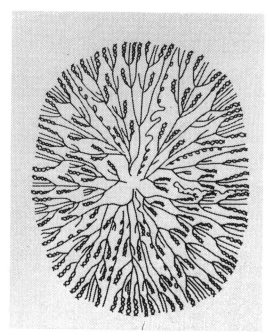

Fig. 3.6 A simplified model of the molecular structure of the starch grain (Lineback, 1984, reproduced with kind permission of the publishers).

density, crystallinity and susceptibility to enzyme and chemical attack. It has been suggested that the thickness of a ring could correspond to the length of the amylose and amylopectin molecules.

Within the starch grain the amylose and amylopectin molecules are probably held together by hydrogen bonds. Commercial 'dry' starch contains 10–17 per cent water, which is not readily removed from starch, and is necessary for the three-dimensional order of the starch granule. In the presence of excess cold water, the starch granule absorbs more water and swells. This swelling is reversible at low temperatures, but becomes irreversible as the temperature is raised, and the hydrogen-bonds which hold the amylose together are broken.

The simplified model of the molecular organisation of the amylopectin and amylose molecules in the starch granule given in Figure 3.6 shows a possible arrangement of these molecules within a single concentric ring of a starch granule. The helical chains of the amylose and amylopectin A-chains probably intertwine.

Granule development

A starch granule grows by apposition. Newly synthesised material is deposited on the surface and the granule grows in size. Generally, as a starch-accumulating organ matures, the average molecular size of both the amylose and amylopectin molecules increases, and the proportion of amylose rises. However, when we try and relate these observations to an understanding of how the starch granule develops, we are faced by the difficulty presented by the developmental and ontogenetic heterogenity of the tissues that synthesise starch. It has not yet proved possible to analyse a single starch grain at different stages of development, and when we sample a developing organ, we sample a variety of tissues which contribute differentially at various stages of maturity to the starch content. For example in a cereal grain, initially it is the pericarp where most of the starch is found, later as starch accumulates in the endosperm the pericarp starch becomes insignificant. Even within a tissue like the endosperm there is no homogeneity, since during development the central region comes to contain a higher concentration of starch than the peripheral regions.

The concentric rings which are apparent under the light microscope in some large granules, such as those of potato (Figure 3.5), suggest that starch is deposited with a diurnal rhythm, just as growth rings in trees arise from the annual growth rhythm of the secondary cambium. These growth rings are produced even when the potatoes are grown in a constant environment, which could mean that there is an endogenous rhythm of deposition. With small granule starches growth rings are seldom seen under the light microscope, but when the granules are eroded by enzyme or chemical attack, concentric rings can be visualised with the electron-microscope. The significance of these growth rings is not well understood. In wheat there is one growth ring per day. These rings disappear when wheat is grown under continuous illumination, and again as in the potato, they may represent the periodic growth of the granule.

Leaving aside the starch granule itself, and considering the development of the plant organ that accumulates the starch, we find an important difference between vegetative organs, in which starch synthesis occurs throughout growth, and reproductive organs, in which starch synthesis occurs in an organ which is virtually fully grown. In vegetative organs, such as stem tubers and roots, meristematic activity is retained throughout the period of growth, and continued growth depends on uninterrupted cell division. In these organs, once initiation has occurred, starch accumulation appears to be a stimulus for cell division and may even be viewed as a driving force for organ growth and development. Thus to take the example of potato tubers, the onset of tuberisation is immediately marked by a rise in the content of starch per cell, and in the contribution of starch to the dry weight.

By contrast, if we take the endosperm of wheat as an example of a reproductive structure, we see a different pattern. Here development starts with a series of divisions of the primary endosperm nuclei followed by the laying down of primary cell walls. Only when this cellularisation is complete (some 14–20 days after anthesis in wheat) does starch accumulation begin, with the central region, the starchy endosperm predominating as the site of starch synthesis. A single starch granule develops within each amyloplast. A pattern similar to that of wheat occurs in barley, oats, rye and rice, but in oats >100 granules occur in each amyloplast and in rice the central endosperm cells contain about 80 granules per amyloplast. In maize there is a single starch granule per amyloplast, but as an important exception to the general rule, cell division continues during the grain filling period. In seeds of starchy legumes, such as the garden pea, starch accumulates only after the cotyledons have reached maximum size.

BIOCHEMICAL PATHWAY OF STARCH BIOSYNTHESIS

The primary substrate for starch synthesis is generally considered to be sucrose. There are a number of reasons for this view. Sucrose is the form in which carbon is translocated in most plants. Cereal endosperm, potato tuber slices and other starch accumulating tissues can be shown to make starch *in vitro* from sucrose supplied to them. Enzymes capable of converting sucrose to the precursors of starch are present in starch-accumulating tissues. The intracellular concentrations of sucrose in the cereal endosperms have been estimated to be up to 90 mM, and in potato tubers up to 20 mM. It is likely that most of this sucrose is localised in the large vacuoles which are a conspicuous feature of cells in the early stages of starch accumulation.

The enzyme responsible for the initial mobilisation of sucrose is probably sucrose synthase. At first this enzyme was considered responsible for sucrose synthesis, but it is now recognised that sucrose synthase is more likely to be responsible for the initial cleavage of sucrose as the first step in the conversion of sucrose to storage and structural carbohydrate. In support of this role for sucrose synthase, it should be noted that the reaction catalysed is freely reversible, and the enzyme is widely distributed in tissues that receive translocated sucrose. Invertase activity is also detectable in these tissues, but at a relatively low level. This enzyme may be responsible for much of the free glucose and fructose observed during the early stages of starch deposition.

The enzyme responsible for the polymerisation of glucose to starch is starch synthase. Most activity is found to be firmly bound to the starch granule, with only a minor soluble fraction, which is presumably located in the stroma of the amyloplast. While the soluble starch synthases are specific for ADP-glucose as the glucosyl donor for polymerisation, the granule-bound starch synthases are able to use either ADP-glucose or UDP-glucose. However, the kinetics of the UDP-glucose-dependent activity speak against a role for this nucleotide sugar *in vivo*: both the velocity of the reaction and the affinity for UDP-glucose compare unfavourably with those observed with ADP-glucose. There is additional evidence from mutant varieties of maize (see below) which points to ADP-glucose as the sole source of glucosyl groups for starch synthesis *in vivo*.

Starch synthesis is confined to the amyloplast. Electron-microscopy reveals that the developing amyloplast resembles the chloroplast in being bounded by a double membrane. What electron-microscopy cannot reveal is the nature of the selective permeability of the amyloplast envelope. Up to the present the transport properties of the isolated amyloplast have not been identified, and the form in which assimilate enters the amyloplast is not known. One reason for this lack of knowledge is that it has proved difficult to isolate intact amyloplasts for transport experiments, although some success has been obtained in isolating amyloplasts with sufficient intactness for their complement of enzymes to be determined. However, in general, it has been found that when attempts are made to isolate intact amyloplasts the membrane envelope breaks, presumably because the relatively dense starch granule bursts through the fragile limiting membranes during the differential sedimentation required to separate the amyloplasts from the other cell organelles. So, until more direct evidence is obtainable, we can make only informed guesses as to the biochemical form in which the products of sucrose mobilisation are transferred into the amyloplast for the synthesis of starch.

Amyloplasts resemble chloroplasts in their development from proplastids. Both organelles contain, in general terms, a similar array of enzymes. Under certain conditions amyloplasts and chloroplasts are interconvertible. Moreover the amyloplast, like the chloroplast, is surrounded by two membranes, which are retained throughout development; in the potato tuber the two membranes are still detectable after 9 months of storage. These similarities between the two organelles have suggested that the amyloplast has a membrane transport system similar to that of the chloroplast. In the chloroplast the outer membrane is permeable to a wide range of metabolites, while the inner membrane bears the carriers that control the flows of metabolities into and out of the chloroplast.

If it were possible for the ADP-glucose to enter the amyloplast, the flow of carbon from sucrose to amylose would involve, in addition to sucrose synthase and starch synthase, only the regeneration of ADP-glucose from cytosolic fructose. However, transport systems for ADP-glucose (or UDP-glucose) are absent from chloroplasts, and have not been found elsewhere in cells. Consequently the possibility of entry as a nucleoside diphosphate glucose is generally ruled out.

In the mature chloroplast triose phosphate and phosphoglyceric acid cross the membrane either in exchange for one another or in exchange for inorganic phosphate. The transport catalysed is an equimolar, electroneutral, reversible exchange. When the chloroplast is actively exporting assimilate, either during active photosynthesis or during the nocturnal breakdown of accumulated starch, triose phosphate flows out of the chloroplast coupled to an inward flow of inorganic phosphate. If a similar carrier were to operate in the developing amyloplast, this exchange would occur in a reverse direction, so that triose phosphate would flow inwards in exchange for an efflux of inorganic phosphate. If assimilate is taken into the amyloplast through this carrier then the pathway taken by assimilate could be as shown in Figure 3.7.

In this pathway, conversion of the UDP-glucose in the cytosol to glucose 1-phosphate can be explained by the activity of the UDP-glucose pyrophosphorylase. High activities of this enzyme have been observed in a range of starch-storing tissues. The glucose 1-phosphate formed would be converted to triose phosphate by the enzymes of the glycolytic sequence. The fructose moiety of sucrose contributes equally to the starch formed. The fructose released from the sucrose would join the flow of assimilate after conversion to fructose 6-phosphate by the action of hexokinase, or the more specific fructokinase (Table 3.2). Within the amyloplast glucose 1-phosphate would be reformed by an effective reversal of the cytosolic glycolysis. The direction of the intra- and extra-amyloplast conversions of hexose and triose phosphates would be controlled by the relative activities of the enzymes that interconvert fructose 6-phosphate and fructose 1,6-bisphosphate: in the cytosol a high activity of an ATP-dependent phosphofructokinase would ensure the formation of fructose 1,6-bisphosphate, and thus of triose phosphate; while in the amyloplast a high activity of the fructose 1,6-bisphosphatase would ensure the formation of fructose 6-phosphate, which could in turn be converted via freely reversible reactions to the starch precursor, ADP-glucose.

Even though it has not been possible to study directly transport into the amyloplast, the correctness of the metabolic pathway shown in Figure 3.7 has not entirely escaped experimental testing. At least two of the requirements of this scheme have recently been examined, and in both cases the results obtained have weakened the case for the pathway shown in Figure 3.7.

First, plant tissues have a high activity of the cytosolic enzyme triose phosphate isomerase, that catalyses the interconversion of dihydroxyacetone phosphate and phosphoglyceraldehyde, and it is assumed that these two isomers are in equilibrium within the cytosol. This isomerisation effectively

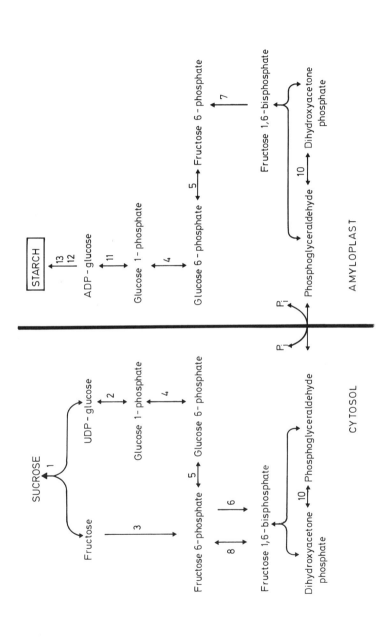

Fig. 3.7 Possible metabolic pathway of starch synthesis in which assimilate enters the amyloplast via a triose phosphate/inorganic phosphate translocator. The enzymes involved are indicated by the numbers as given in Table 3.2.

Table 3.2 Enzymes probably involved in the synthesis of starch from sucrose.

---

1. Sucrose synthase
   UDP-glucose:D-fructose 2-α-D-glucosyltransferase   EC 2.4.1.13

   $$\text{UDP-glucose} + \text{fructose} \leftrightarrow \text{UDP} + \text{sucrose}$$

2. UDP-glucose pyrophosphorylase
   UTP:α-D-glucose 1-phosphate uridylyltransferase   EC 2.7.7.9

   $$\text{UTP} + \text{glucose 1-phosphate} \leftrightarrow \text{UDP-glucose} + \text{PP}_i$$

3. Fructokinase
   UTP:D-fructose 6-phosphotransferase   EC 2.7.1.4

   $$\text{UTP} + \text{fructose} \rightarrow \text{UDP} + \text{fructose 6-phosphate}$$

4. Phosphoglucomutase
   α-D-glucose 1,6-bisphosphate:α-D-glucose 1-phosphate phosphotransferase
                                                                       EC 5.4.2.2

   $$\text{glucose 6-phosphate} \leftrightarrow \text{glucose 1-phosphate}$$

5. Phosphoglucose isomerase
   D-glucose 6-phosphate ketolisomerase   EC 5.3.1.9

   $$\text{glucose 6-phosphate} \leftrightarrow \text{fructose 6-phosphate}$$

6. ATP-dependent phosphofructokinase
   D-fructose 6-phosphate 1-phosphotransferase   EC 2.7.1.11

   $$\text{ATP} + \text{fructose 6-phosphate} \rightarrow \text{ADP} + \text{fructose 1,6-bisphosphate}$$

7. Fructose 1,6-bisphosphatase
   D-fructose 1,6-bisphosphate 1-phosphohydrolase   EC 3.1.3.11

   $$\text{fructose 1,6-bisphosphate} + \text{H}_2\text{O} \rightarrow \text{fructose 6-phosphate} + \text{P}_i$$

8. Pyrophosphate-dependent phosphofructokinase
   Pyrophosphate:D-fructose 6-phosphate 1-phosphotransferase   EC 2.1.7.90

   $$\text{PP}_i + \text{fructose 6-phosphate} \leftrightarrow \text{P}_i + \text{fructose 1,6-bisphosphate}$$

9. Aldolase
   D-fructose 1,6-bisphosphate-D-glyceraldehyde 3-phosphate-lyase   EC 4.1.2.13

   $$\text{fructose 1,6-bisphosphate} \leftrightarrow \text{glyceraldehyde 3-phosphate} + \text{dihydroxyacetone phosphate}$$

10. Triose phosphate isomerase
    D-glyceraldehyde 3-phosphate ketolisomerase   EC 5.3.1.1

    $$\text{glyceraldehyde 3-phosphate} \leftrightarrow \text{dihydroxyacetone phosphate}$$

11. ADP-glucose pyrophosphorylase
    ATP:α-D-glucose 1-phosphate adenyltransferase   EC 2.7.7.27

    $$\text{ATP} + \text{glucose 1-phosphate} \leftrightarrow \text{PP}_i + \text{ADP-glucose}$$

Table 3.2 (*continued*)

---

12. Starch synthase
    ADP(UDP)glucose:α-1,4-D-glucan α-4-D-glucosyltransferase   EC 2.4.1.21

    $$\text{ADP(UDP)glucose} + (\text{glucosyl})_n \rightarrow \text{ADP(UDP)} + (\text{glucosyl})_{n+1}$$

13. Branching enzyme
    α-1,4-D-glucan:α-1,4-glucan α-6-D-glucanotransferase   EC 2.4.1.18

    $$\text{1,4 glucan} \rightarrow \text{1,6 glucosyl 1,4 glucan}$$

14. Starch phosphorylase
    α-1,4-D-glucan:orthophosphate α-glucosyltransferase   EC 2.4.1.1

    $$\text{glucose 1-phosphate} + (\text{glucosyl})_n \leftrightarrow P_i + (\text{glucosyl})_{n+1}$$

---

allows an equilibration between the two halves of the fructose 1,6-bisphosphate molecule, and thus of the hexoses which are derived from it. On the basis of this assumption, it can be predicted that if triose phosphates are necessary intermediates in the metabolic pathway from sucrose to starch, then the glucosyl residues of starch synthesised from hexone specifically labelled in the carbon-1 or carbon-6 positions should not retain the asymmetrical labelling, but have the label redistributed between the carbon-1 and carbon-6 positions. Conversely if triose phosphates are not directly involved in the pathway, and starch is made from hexoses derived from sucrose without cleavage into triose, then the glucosyl residues of the starch should retain the asymmetrical labelling of the substrate hexose. This was the rationale of the experiments reported by Keeling *et al.* (1988). They found that when they fed to endosperm of developing wheat grains glucose or fructose labelled with the ^{13}C either in the carbon-1 or carbon-6 position, the asymmetric distribution of label was largely retained in the glucose released from the newly synthesised starch. Evidence was provided to show that the limited redistribution of label, which occurred to an extent of 15–25 per cent, was consistent with a proportion of the assimilate flowing through the cytosolic pool of triose phosphate (Figure 3.8). They concluded that the flow of hexose units from sucrose to starch did not necessitate conversion to triose phosphate for transport into the amyloplast, which they suggested takes place as a hexose monophosphate such as fructose 6-phosphate, glucose 6-phosphate or glucose 1-phosphate (Figure 3.8).

The second piece of evidence to cast doubt on the role of triosephosphate as the actual substrate for starch synthesis within the amyloplast comes from a careful analysis by Entwistle and ap Rees (1988) of the enzyme complement of intact amyloplasts isolated from wheat endosperm. They found that the enzymes of starch synthesis and of the glycolytic pathway were well represented in the amyloplasts, but fructose 1,6-bisphosphatase activity was

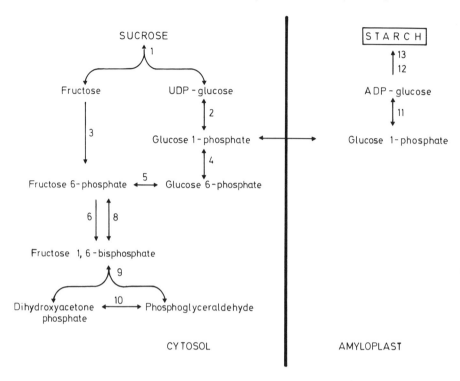

Fig. 3.8   A metabolic pathway for the synthesis of starch from sucrose in developing wheat endosperm, showing transport of hexose monophosphates across the amyloplast envelope. The enzymes involved are indicated by the numbers given in Table 3.2.

absent. Moreover in endosperm homogenates no fructose 1,6-bisphosphatase activity was detected and no fructose 1,6-bisphosphatase was recognised by an antibody prepared against the chloroplast enzyme. Fructose 1,6-bisphosphate can be converted to fructose 6-phosphate by the activity of pyrophosphate-dependent phosphofructokinase (Table 3.2), but in wheat endosperm this enzyme was found to be confined to the cytosol. Thus the additional absence of fructose 1,6-bisphosphatase from the amyloplasts would mean that these organelles lack the capability to convert fructose 1,6-bisphosphate to fructose 6-phosphate, and would thus be unable to use triose phosphate for starch synthesis (Figure 3.7).

Taking together these two lines of evidence it may, therefore, be appropriate to modify the metabolic pathway given above (Figure 3.7) to show the flow of assimilate entering the amyloplast as hexose rather than triose units (Figure 3.8).

Immature chloroplasts contain starch made from imported assimilate. When they make starch from exogenous sucrose they act very much like amyloplasts. There is some evidence that the hexose units which constitute the starch made under these conditions have not been broken down to triose phosphate during starch synthesis, and therefore it is likely that entry to the chloroplast, when that organelle is immature, occurs as hexose phosphate. Thus the proper model for the developing amyloplast may well be the developing, and not the mature, chloroplast.

However, it needs to be emphasised that any conclusions on the metabolic pathway of starch synthesis must remain tentative until the technical problems of amyloplast isolation have been solved, so that we can have more direct information on the transport capabilities of the amyloplast envelope during the development of the starch grains. Also we need to see the focus of the work widened from its present concentration on the wheat endosperm to encompass a greater variety of starch-storing systems.

Whatever the form in which assimilate enters the amyloplast, the glucosyl donor for starch synthesis is likely to be ADP-glucose. Synthesis of ADP-glucose from glucose 1-phosphate can be accounted for by the activity of the ADP-glucose pyrophosphorylase. The enzyme from potato tubers and maize endosperm resembles the well-characterised spinach leaf enzyme with a M_r of 200 kDa and a tetrameric subunit structure, with two types of subunit. Activity of the chloroplast ADP-glucose pyrophosphorylase is stimulated by phosphoglyceric acid and inhibited by inorganic phosphate. These effects can account for the activation of the enzyme in the transition from dark to light, when phosphoglyceric acid levels rise and inorganic phosphate levels fall. The enzyme isolated from maize endosperm, potato and cassava is similarly stimulated by phosphoglyceric acid and inhibited by inorganic phosphate. When the enzyme from maize endosperm was first isolated and studied *in vitro* it appeared to be quite different from the leaf and potato enzymes, but these differences were later shown to be attributable to protease action during the isolation. It now seems that both the photosynthetic and non-photosynthetic ADP-glucose pyrophosphorylases resemble one another in their structure and sensitivity to phosphoglyceric acid and inorganic phosphate. But what can the significance of these effects of phosphoglyceric acid and inorganic phosphate be for starch synthesis in storage tissues? Unlike the mature photosynthetic tissues, storage tissues are undergoing a rapid development accompanying starch synthesis. Thus while the synthesis of starch in photosynthetic tissues responds to light via changes in the level of regulatory metabolites, it is likely that starch synthesis in storage tissues is controlled more by the regulation of the synthesis and breakdown of the biosynthetic enzymes.

In catalysing the transfer of glucose residues to the non-reducing ends of primer molecules, starch synthase leads to the formation of linear chains of α-1,4 glucan. However, it is unable to create new polysaccharide chains *de novo*. The nature and source of the primer molecules *in vivo* are unknown. Oligosaccharides of the series maltose, maltotriose can be used *in vitro*, as well as larger molecules of amylose and amylopectin. *In vivo* the oligosaccharide primers could be synthesised by the activity of starch phosphorylase or of enzymes yet unknown. Starch phosphorylase is present in tissues actively synthesising starch. At one time it was thought to be responsible for the polymerisation of glucose units to starch. However, the relative concentrations of inorganic phosphate and glucose 1-phosphate likely to be present in the amyloplasts, and a high K_m for glucose 1-phosphate both favour the phosphorolytic breakdown of starch by starch phosphorylase rather than its synthesis by this enzyme.

The linear chains generated by the activity of the starch synthase would provide only the amylose fraction of the starch granule. The amylopectin fraction, which usually constitutes some three quarters of the starch, is formed by the introduction of α-1,6 linkages into the growing, linear molecules by the action of the branching enzyme.

Activity of the branching enzyme can be assayed by measuring the decrease in absorbance of the amylose-iodine complexes, as amylose is converted to amylopectin. Alternatively it can be measured by determining the stimulation of D-glucan synthesis from glucose 1-phosphate catalysed by starch phosphorylase. The branching enzyme stimulates this activity by increasing the number of sites for further elongation by the starch phosphorylase. The latter method of assay may represent the process *in vivo* where new non-reducing ends are available to the starch synthase enzyme for elongation.

In a variety of plants multiple forms of both soluble and granule-bound starch synthases and of the branching enzyme have been separated by DEAE-cellulose column chromatography. In maize the two forms of soluble starch synthase, designated starch synthase I and II, differ in their affinity for primers and in their relative activities assayed *in vitro*. Digestion of the starch granule with α-amylase and amyloglucosidase releases the granule-bound starch synthase activity, and once again two forms (designated I and II) resembling the two soluble enzymes have been separated. When the granule-bound activity is released from the starch granules, it loses the ability to use UDP-glucose, and thus comes to resemble the soluble enzyme in its specific use of ADP-glucose. This finding raises the question as to what extent the soluble and granule-bound starch synthases are actually different. An answer is beginning to become available. Both the soluble starch synthase II and the granule-bound starch synthase II have a M_r of about 93 kDa. However, while the soluble starch synthase I has a M_r of about 70 kDa, the granule-bound starch synthase I is significantly smaller with a M_r of about 60 kDa. The granule-bound and soluble starch synthase I isoforms also differ in kinetic characteristics, and in their immunological cross-reactivity, and therefore may well be quite distinct enzymes.

In order to understand how starch is made, it is obviously crucial to discover how these different forms of starch synthase and branching enzyme interact to construct the amylopectin. Probably they form a complex and work in concert, rather than taking part in a sequential process in which amylose is made first and then branches are introduced to form amylopectin. At the early stages of granule development amylopectin synthesis predominates, with the proportion of amylopectin decreasing as the starch is deposited.

The effect of the branching enzyme is to generate more non-reducing ends of polymers as substrates for the starch synthase. As the chains elongate by the activity of the starch synthase, new branches are introduced by the branching enzyme. Possibly the different forms of starch synthase and branching enzyme have different substrate specificities, and consequently are responsible for different chain lengths and different branching patterns.

This simple concept of starch synthesis leaves many important problems unsolved. How does the amylose escape from the branching activity of the branching enzyme? How are the branching and polymerisation reactions in amylopectin synthesis controlled so that the particular primary structure is arrived at? How is the final shape and size of the starch granule determined? Clearly progress in solving these problems will come when the multiple forms of the starch synthase and branching enzyme have been isolated in a pure form uncontaminated by other enzymes. Only then will their substrate specificities and the products of their catalytic activities be identified. The importance of this information for the biotechnological utilisation of starch becomes apparent when we realise that it is the results of the interaction of these multiple forms of starch synthase and branching enzyme which directly affect the quality of the product, since the characteristics of the starch in terms of texture, viscosity and stability are all strongly influenced by the relative proportions and molecular size of the amylose and amylopectin fractions. An indication of how the biosynthesis of starch is affected by deficiencies in certain enzymes, and enzyme isoforms has come from studies of the enzyme deficiencies associated with the aberrant starch of mutant varieties of maize and pea.

MUTANTS OF STARCH SYNTHESIS

In maize, mutations affecting the filling of the grain are easily recognised by the altered appearance of the grain. The particular designation of the mutant variety indicates its phenotypic appearance, for example *brittle* and *dull*. The mutant variety *waxy* acquired its name from the shiny appearance of the cut surface of the endosperm. It looked as though it contained wax, although of course the appearance was due not to wax but to an altered composition of the starch. This variety first became important when a replacement for tapioca (obtained from cassava) was sought in the USA during World War II. Since then, *waxy* maize has continued to be grown extensively because of the particular properties of the starch produced. Some varieties have been developed from a deliberate breeding programme. For example the high amylose varieties resulted from a breeding programme which set out to increase the proportion of the linear polymer, when it had been recognised that amylose, like other natural straight-chain polymers such as cellulose, could form films and fibres with potential industrial applications.

Usually the endosperm of the mutant varieties is smaller than normal and contains less starch. In some the ratio of amylose to amylopectin is altered. Thus *shrunken-1* has a lower proportion of amylose, and *amylose extender* a

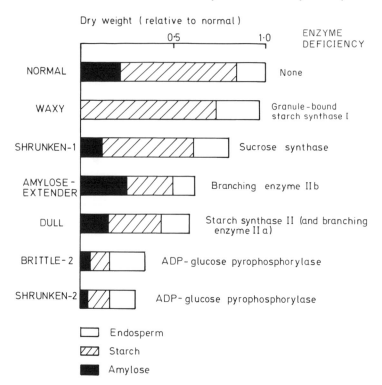

Fig. 3.9   Relative size and starch and amylose compositions of the mature endosperms of various genotypes of maize, in which a specific enzyme deficiency has been identified (data from Nelson, 1980).

greater proportion of amylose than normal. *Waxy* has no amylose at all. *Sugary-1* has a normal proportion of amylose but almost 40 per cent of the dry weight is made up of a water-soluble polysaccharide not found in normal mature endosperm. This is called phytoglycogen from its resemblance to animal glycogen. It is highly branched and has a high M_r.

The size and shape of the mature starch granules are most altered with the *amylose extender* mutation, thus showing the importance of the amylopectin component for granule form. In the developing grain most mutations lead to increased concentrations of sucrose and reducing sugars, which build up as a result of the decreased rate of starch synthesis. In many of the mutant endosperms enzyme deficiencies that lead to the particular altered condition have been identified (Figure 3.9). Enzymes that catalyse similar reactions in other parts of the plant, that is, the sporophytic tissues, are unaffected by the mutations. Thus for example the starch in the embryos of *waxy* seeds, unlike that in the endosperm, stains blue indicating the presence of amylose.

In *amylose extender* the mutation affects the level of one of the isoforms of the branching enzyme. The other forms of the branching enzyme and the starch synthase are unaffected. This is consistent with the observed low percentage of amylopectin and the high average chain length of the amylopectin that is present.

Dull is another mutation associated with an increased proportion of amylose and a decreased content of starch. However, in this case chromatographic separation of the different forms of starch synthase and branching enzyme reveals a lower level of one form of starch synthase and a small decrease in one of the branching enzymes. Because two enzymes are affected in *dull*, it is thought that the mutation may be affecting a regulatory gene rather than a structural gene.

There is now good evidence to indicate that the *waxy* locus is the structural gene for the granule-bound, ADP-glucose-dependent, starch synthase I isoform, and that this enzyme is responsible for amylose synthesis. This is a most important conclusion, since it allows us to ascribe a specific role to a particular isoform of starch synthase. It also indicates the extent to which the nature of the starch produced by a crop plant can be altered by a relatively minor genetic alteration. In the case of the *waxy* variety the mutation arose spontaneously. In the future as our understanding of the biochemical basis of starch synthesis is increased then it should become possible by genetic manipulation to tailor the starch produced by a crop such as maize to the demands of the end user, be it a food processor or a manufacturer of plastics.

FURTHER READING

Badenhuizen, N.P. (1969). *The Biosynthesis of Starch Granules in Higher Plants*. New York, Appleton-Century-Crofts.

Banks, W. & Muir, D.D. (1980). Structure and chemistry of the starch granule, in *The Biochemistry of Plants. Vol. 3 Carbohydrates: Structure and Function*, Ed Preiss, J., New York, Academic Press, pp. 321–369.

Blanshard, J.M.V. (1987). Starch granule structure and function: a physicochemical approach, in *Starch: Properties and Potential. Critical Reports on Applied Chemistry. Vol. 13*, Galliard, T., Chichester, John Wiley, pp. 16–54.

Duffus, C.M. & Cochrane, M.P. (1982). Carbohydrate metabolism during cereal grain development, in *The Physiology and Biochemistry of Seed Development, Dormancy, and Germination*, Ed Khan, A.A., Amsterdam, Elsevier/North Holland, pp. 43–66.

Duffus, C.M. & Duffus, J.H. (1984). *Carbohydrate Metabolism in Plants*. London, Longmans.

French, D. (1984). Organization of starch granules, in *Starch: Chemistry and Technology*, Eds Whistler, R.L., Bemiller, J.H. & Paschall, E.F., New York, Academic Press, pp. 184–247.

Galliard, T. (1987). Starch availability and utilization, in *Starch: Properties and Potential. Critical Reports on Applied Chemistry. Vol. 13*, Ed Galliard, T., Chichester, John Wiley, pp. 1–15.

Jenner, C.F. (1982). Storage of starch, in *Plant Carbohydrates I Intracellular Carbohydrates. Encyclopaedia of Plant Physiology Vol. 13A*, Eds Loewus, F.A. & Tanner, W., Berlin, Springer Verlag, pp. 700–747.

Jennings, J.L. (1987). Starch crops, in *Handbook of Plant Science in Agriculture Vol. II*, Ed Christie, B.R., Boca Raton, Florida, CRC Press, pp. 137–143.

Kainuma, K. (1988). Structure and Chemistry of the Starch Granule, in *The Biochemistry of Plants Vol. 14*, Ed Preiss, J., New York, Academic Press, pp. 141–180.

Kennedy, J.F. Cabral, J.M.S., Sa-Correia, I. & White, C.A. (1987). Starch biomass: a chemical feedstock for enzyme and fermentation processes, in *Starch: Properties & Potential. Critical Reports on Applied Chemistry. Vol. 13*, Ed Galliard, T., Chichester, John Wiley, pp. 115–148.

Koch, H. & Röper, H. (1988). New industrial products from starch, *Starch/Stärke* **40**, 121–131.

Mares, D.J., Sowokinos, J.R. & Hawker, J.S. (1985). Carbohydrate metabolism in developing potato tubers, in *Potato Physiology*, Ed Li, P.H., New York, Academic Press, pp. 279–327.

Munck, L., Rexen, F. & Valby, H.L. (1988). Cereal starches within the European Community—agricultural production, dry and wet milling and potential use in industry, *Starch/Stärke* **40**, 81–87.

Otey, F.H. & Doane, W.M. (1984). Chemicals from starch, in *Starch: Chemistry and Technology*, Ed Whistler R.L., Bemiller, J.N. & Paschall, E.F., New York, Academic Press, pp. 389–468.

Preiss, J. (1988). Biosynthesis of starch and its regulation, in *The Biochemistry of Plants Vol. 14*, Ed Preiss, J., New York, Academic Press, pp. 181–254.

Preiss, J. & Levi, C. (1980). Starch biosynthesis and degradation, in *The Biochemistry of Plants. Vol. 3 Carbohydrates: Structure and Function*, Ed Preiss J., New York, Academic Press, pp. 371–423.

Shannon, J.C. & Garwood, D.L. (1984). Genetics and physiology of starch development, in *Starch: Chemistry and Technology*, Eds Whistler, R.L., Bemiller, J.N. & Paschall, E.F., New York, Academic Press, pp. 26–86.

Zobel, H.F. (1988). Molecules to granules: a comprehensive starch review, *Starch/Stärke* **40**, 44–50.

ADDITIONAL REFERENCES

Commission of the European Communities (1984). *The Production and Use of Cereal and Potato Starch in the EEC.* Centre for European Agricultural Studies (Wye College, University of London) and Institut de Gestion Internationale Agro-Alimentaire (CERGY-Pontoise).

Entwistle, G. & ap Rees, T. (1988). Enzymic capacities of amyloplasts from wheat (*Triticum aestivum*) endosperm, *Biochem. J.* **255**, 391–396.

Keeling, P.L., Wood, J.R., Tyson, R.H. & Bridges, I.G. (1988). Starch biosynthesis in developing wheat grain. Evidence against the direct involvement of triose phosphates in the metabolic pathway, *Plant Physiol.* **87**, 311–319.

Lineback, D.R. (1984). The starch granule organisation and properties, *Bakers Digest* **58**, 16–21.

Nelson, O.E. (1980). Genetic control of polysaccharide and storage protein synthesis in the endosperms of barley, maize and sorghum, in *Advances in Cereal Science and Technology. Vol. 3*. Ed Pomeranz, Y., Saint Paul-Minnesota, American Association of Cereal Chemists, pp. 41–71.

Robin, J.P., Mercier, C., Charbonnier, R. & Guilbot, A. (1974). Lintnerized starches. Gel filtration and enzymatic studies of insoluble residues from prolonged acid treatment of potato starch, *Cereal Chem.* **51**, 389–406.

Chapter 4

Fructan

INTRODUCTION

Fructose is the sweetest naturally occurring sugar. Depending on how the tests are carried out, fructose can taste from 20 to 80 per cent sweeter than sucrose. In addition, fructose is more soluble than sucrose, less viscous, causes fewer dental caries, and low levels can be metabolised without a need for insulin. For the food and soft drinks industries fructose possesses so many organoleptic and technical advantages over sucrose that when a cheap source of fructose, high-fructose corn syrup (HFCS), was introduced in the USA in the 1970s, the sweetener industry was revolutionised. By 1985 high-fructose corn syrup was accounting for 30 per cent of the sweetener market in the USA, compared with the 43 per cent market share of sucrose, the remainder being taken by glucose syrups and artifical 'high intensity' sweeteners (Fuchs, 1987). It is predicted that by the year 2005, high-fructose corn syrup and crystalline fructose will together have overtaken sucrose with a total annual production in the USA equivalent to 6.6 million tonnes of sucrose (Fuchs, 1987).

As its name suggests, high-fructose corn syrup is produced from maize (USA, corn) starch, which is first hydrolysed enzymatically, and the resulting glucose is isomerised, again enzymatically, to give a syrup containing 42 per cent fructose and 50 per cent glucose. The fructose content is upgraded to 90 per cent by chromatographic separation of fructose from glucose. Largely for economic and political reasons these technical developments have been confined to the USA. For example, in the European Community the production and utilisation of high-fructose corn syrup has been restricted in order to protect the interests of the beet-sugar farmers and the associated industry. However, in the face of the technical advantages of fructose over sucrose this

56

Table 4.1 Biotechnological utilisation of fructan (modified from Fuchs, 1987).

Process	Product	Use
Food		
Enzymatic hydrolysis of purified fructan	High-fructose syrup	Soft drinks and food
	Crystalline fructose	
Fermentation of unpurified juices for bulk chemicals		
Kluyveromyces marxianus (yeast)	Bio-ethanol	Fuel
Clostridium acetobutylicum (bacterium)	Acetone, butanol	Solvents
Bacillus polymyxa (bacterium)	2,3-butanediol	Fuel additive
Synthesis of fine chemicals		
Hydrogenolysis	Glycerol	
	1,2-propanediol	
	Glycol	
Hydrolysis and dehydration	Hydroxymethylfurfural	Furan chemicals
Hydrolysis and hydrogenation	Mannitol	Low calorie sweetener

policy is unlikely to continue indefinitely, and in the future fructose will probably increase in importance throughout the world.

Cereal starch that has been hydrolysed and isomerised is not the only source of fructose, although it is of course the most economical. In some respects a more natural alternative is the fructose polymer, fructan. This reserve carbohydrate of many plants is accumulated to high levels in storage organs of species belonging to the Compositae, notably Jerusalem artichoke and chicory. Both species accumulate carbohydrate to levels comparable to sugar beet and potato (Figure 2.2), and both can be readily cultivated and harvested. The fructose is released on hydrolysis of the fructan. But as well as being a source of fructose, the fructan itself is slowly finding applications in the food industry. Fructans have about 10 per cent of the sweetness of a comparable weight of sucrose, and possess many of the technical properties of sucrose required by the food industry (Chapter 2). However, when ingested, fructans remain unhydrolysed until they reach the large intestine where they encourage the selective growth of bifidobacteria which are claimed to benefit health. Because the fructan is less readily absorbed from the digestive tract it is classed as a low-calorie, natural, dietary fibre.

As well as the applications in the food industry, the high yields of fermentable carbohydrate make fructan-derived fructose suitable as a bulk chemical in the production of ethanol, acetone + butanol, and 2,3-butanediol by apropriate one-step fermentations (Table 4.1). Additionally, as a fine chemical fructose has advantages over glucose, derived from starch or cellulose, in the production of a variety of furan chemicals (Table 4.1).

The study of fructan has long been neglected in comparison to starch and cellulose, perhaps because of its former commercial unimportance. Thus in examining fructan structure and biosynthesis we shall be relying on a relatively small amount of experimental work, much of which was carried out many years ago without the techniques that are available now.

FRUCTAN STRUCTURE

Fructans are polymers of D-fructose, but each chain contains a single D-glucosyl residue attached by a β-2,1 linkage as in sucrose, and almost invariably located at the end of each chain. Thus fructan chains form a homologous series of oligomers and polymers which can be considered to be fructosyl derivatives of sucrose. The simplest fructan is the trisaccharide, fructosylsucrose, of which three isomers have been isolated: 1-kestose (isokestose), 6-kestose (kestose), and neokestose (Figure 4.1).

The three isomers form the basis of the three types of fructan (Table 4.2). Both 1-kestose and neokestose are elongated by the addition of fructosyl units linked via β-2,1 linkages, while 6-kestose is elongated to form the fructans of grasses via the addition of β-2,6-linked fructosyl units.

The neokestose series of fructans differ from both the inulins and phleins in that elongation occurs from both fructosyl residues of the neokestose (Figure

(a)

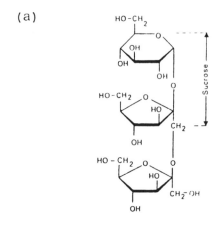

(b)

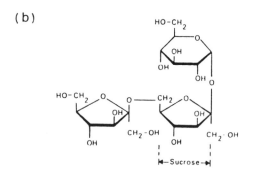

(c)

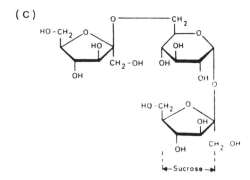

Fig. 4.1 The structures of three isomers of trisaccharide fructan found in plants (a) 1-kestose (isokestose), (b) 6-kestose (kestose), and (c) neokestose (from Pollock & Chatterton, 1988. Reproduced with kind permission of the authors and Academic Press, Inc.).

Table 4.2 The different types of fructan.

Trivial name	Basic trisaccharide	Fructosyl link in elongating chain	Maximum chain length (DP)	Distribution
Inulin	1-kestose	β-2,1	35–50	Compositae Boraginaceae Iridaceae
Phlein	6-kestose	β-2,6	250	Gramineae
	Neokestose	β-2,1	8–10	*Asparagus*

DP, degree of polymerisation.

4.1), so that the glucose residues are found within the chain rather than at the end as in inulins and phleins. These three series of fructans all form linear chains, but branched chains, and chains containing different linkages are also known. Fortunately we can put aside these complications, since for our present purposes we can confine our attention to the simple, linear β-2,1 linked inulins, as these form the only fructans accumulated by the Compositae species under consideration as potential crops. The fructan of chicory and Jerusalem artichoke is not only the most promising in terms of biotechnological exploitation, but the Jerusalem artichoke fructan is also the best characterised in terms of the physiology and biochemistry of synthesis.

The terms inulin and phlein originated with the discovery of fructans in the 19th century: inulin from *Inula helenium*, phlein from *Phleum pratense*. Since that time our greater knowledge of fructan structure and composition has meant that these terms have become imprecise. For this reason the term fructan is gradually replacing the terms inulin and phlein, with the plant source defining the type of specific fructan, for example, chicory fructan. The term fructan has also replaced, among others, the older terms fructosan, sucrosylfructan and polyfructosan, in order to be consistent with the terminology applied to the other polysaccharides, such as glucan for the glucose polymers. Glucofructan, however, would appear to have the merit of greater precision in that it recognises the obligatory presence of the glucosyl residue. In the absence of a strict convention, fructans with a degree of polymerisation (DP) <10 are described as oligomeric, and fructans with a DP >10, as polymeric. For many years this distinction was useful in practical terms, since DP 10 was the upper limit at which individual chains could be resolved by chromatographic techniques. Now with the application of anion exchange chromatography to the analysis of fructans it is possible to resolve the full range both oligomeric and polymeric fructans present in the Compositae (Figure 4.2).

Fructan resembles starch in being a polymeric carbohydrate used by plants as both a short-term and long-term store of assimilate. Yet, despite this functional similarity, fructan and starch have little in common (Table 4.3).

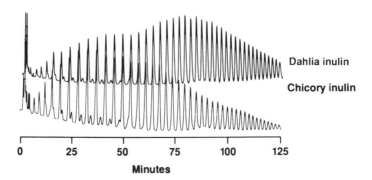

Fig. 4.2 Fructans of dahlia and Jerusalem artichoke tubers separated by anion exchange chromatography (reproduced with permission of Dionex Corp.).

The five-membered furanose rings which make up the fructan polymers give fructans a more flexible structure than do the six-membered pyranose rings of starch. In the Jerusalem artichoke fructans, additional flexibility comes from the exclusion of the ring carbons from the structural backbone of the molecule (Figure 4.1), which takes the form of a loose helix with 4–5 fructosyl units per turn (Phelps, 1965). These structural features help to explain the high solubility of fructans in water.

Additional of ethanol to solutions of fructans causes them to precipitate. This technique has allowed the fructan to be visualised in thin slices of plant tissues as sphero-crystals located in the plant vacuole. Given the high proportion of the cell volume taken up by the vacuole, and the relatively large amounts of fructan accumulated—80 per cent of the dry weight in Jerusalem artichoke tubers—then the vacuolar location of at least the bulk of the stored fructan cannot be seriously questioned.

The differences between fructan and starch with respect to molecular structure, physical properties, and sub-cellular location can be extended even

Table 4.3 Comparison of starch and fructan (modified from Hendry, 1987).

Characteristic	Fructan	Starch
Monomer	Predominantly fructose	Glucose
Ring structure	Flexible-furanose	Rigid-pyranose
Water solubility	High	Low/insoluble
DP	3–250	$1500–3 \times 10^6$
Storage location	Vacuole	Plastid

further when the biosynthetic pathways are compared. Thus these two carbohydrates polymers, with similar functions as short- and long-term carbohydrate reserves, have little in common in terms of their physical properties and biochemistry. What then are the particular conditions that favour one over the other? For some time fructan was viewed as conferring an advantage in terms of cold tolerance. This view was based on the use of fructan by the 'Northern' types of forage grains and cereals, in contrast to the preponderance of starch and sucrose as reserves in the more 'Southern' species with a tropical and subtropical distribution. However, when the range of species examined for their dependence on either fructan or starch is extended beyond the grasses there are many exceptions to this rule (Hendry, 1987). Moreover, the physiological significance of the accumulated fructan in terms of tolerance to cold and freezing is still unclear (Hendry, 1987).

FRUCTAN ACCUMULATION IN JERUSALEM ARTICHOKE

The Jerusalem artichoke is a native of North America, related to the sunflower with which it can hybridise. It is a herbaceous perennial, which each spring rapidly develops a shoot system up to 3 m high. As befits a plant of temperate latitudes, its development is governed by photoperiod. Stem extension depends on long days, while the tubers, which develop from swollen underground stems (stolons), are induced by the shorter days of late summer (< 13.5 hours daylength). As a result the greatest tuber growth occurs in the autumn when the leaves are rapidly senescing; the tubers are filled by assimilate that has accumulated in the stems, the loss of which corresponds with the accumulation of fructan by the tubers (Incoll & Neales, 1970). Thus the pattern of tuber-filling in the Jerusalem artichoke is in marked contrast to that shown by other crops, such as the potato tuber and cereal grain, where current leaf photosynthate is transferred directly to the developing storage organ, and only a small proportion of assimilate is derived by redistribution from the stem or other temporary stores. However, the Jerusalem artichoke can be made to resemble these crops under experimentally controlled conditions: plants are grown for a limited period of long days, then tubers are induced by exposing the plants to an extensive period of short days. Under these conditions stem storage is by-passed, and assimilate is directed to the tubers from the leaves with only a temporary store of sucrose in the stems (Edelman & Jefford, 1968).

During the period of tuber development the flow of sucrose from the stem exerts a powerful force which drives fructan biosynthesis and tuber enlargement. This force is revealed when the normal capacity of tuber development to absorb the downward flux of sucrose is abolished by removing the tuber initials. As a result, the roots, which are normally fibrous, swell into large root tubers filled with fructan (Edelman & Jefford, 1968). In order to explain how this apparently irresistible downward flow of sucrose is absorbed by fructan biosynthesis we need to examine the biochemistry of fructan synthesis.

BIOCHEMICAL PATHWAY OF FRUCTAN BIOSYNTHESIS

It was noted above that fructans differ structurally from polymers of glucose. Fructans also differ fundamentally in their biosynthesis. Unlike the other plant carbohydrates—sucrose, starch and cellulose (and hemicelluloses)—fructan is synthesised without the direct participation of sugar nucleotides. Thus, while the polymeric form of starch results from the successive transfer of glucosyl residues from ADP-glucose (Chapter 3), the polymeric form of fructan results from a successive transfer of fructosyl residues by direct transfructosylation from chains of lower DP. Our current knowledge of the biochemistry of fructan synthesis is based largely on the studies of Edelman and coworkers, who published a series of papers on fructan metabolism in Jerusalem artichoke tubers over the period 1951–1968. They identified the principal enzymes involved in fructan synthesis, depolymerisation and mobilisation at the different stages of tuber development. However, this foundation of knowledge has not been built upon substantially in subsequent years, so that we know little about the properties of these enzymes as proteins and nothing of their molecular structure or how their activities are regulated at the biochemical and genetic levels.

The substrate for fructan synthesis is undoubtedly sucrose. This can be demonstrated *in vivo* by supplying tissues with radioactively labelled sucrose, and following the pattern of incorporation of label into fructan, and also *in vitro*, by using tissue extracts which can synthesise fructan from sucrose. The enzyme responsible for the initial step in fructan synthesis is sucrose-sucrose 1-fructosyl transferase (SST). This enzyme catalyses the transfer of a fructosyl residue from one molecule of sucrose to another with the release of glucose and the synthesis of the trisaccharide 1-kestose (Table 4.4).

The SST-catalysed reaction is irreversible in the direction of 1-kestose synthesis, since the enzyme is unable to catalyse the synthesis of sucrose with glucose as the fructosyl acceptor and 1-kestose as the fructosyl donor. The standard free energy of the glycosidic bond in sucrose is -30 kJ/mol, which is close to that of the nucleotide sugars, such as UDP-glucose (-32 KJ/mol) and significantly higher than that of many other oligo-and polysaccharides. Thus thermodynamically the SST-catalysed reaction is favoured in the direction of 1-kestose synthesis. But there appears to be a kinetic barrier presented by the low affinity which SST shows for sucrose *in vitro*. When K_m values have been determined for the SST enzymes isolated from a variety of fructan-accumulating species, including Jerusalem artichoke, they have all been around 0.1 M. It is possible that the conditions *in vitro* have led to the K_m being seriously overestimated, perhaps because some essential cofactor was lacking *in vitro*. Alternatively, the SST may have high K_m values *in vivo* so that a relatively high concentration of sucrose is required at the intracellular site of SST activity.

Table 4.4 Enzymes involved in fructan synthesis and breakdown in tubers of Jerusalem artichoke.

---

1. Sucrose:sucrose fructosyl transferase (SST)
   Sucrose:sucrose 1^F-β-D-fructosyltransferase   EC 2.4.1.99

   sucrose + sucrose → 1-kestose + glucose

2. Fructan:fructan fructosyl transferase (FFT)
   1,2-β-D-fructan: 1,2-β-D-fructan 1^F-β-D-fructosyltransferase   EC 2.4.1.100

   $$\text{G-F-(F)}_n + \text{G-F-(F)}_m \leftrightarrow \text{G-F-(F)}_{n-1} + \text{G-F-(F)}_{m+1}$$
   donor   acceptor   polymerising fructan
   [G = glucosyl; F = fructosyl; n,m = 1 or > 1]

3. Hexokinase
   ATP-D-hexose 6-phosphotransferase   EC 2.7.1.1

   ATP + hexose → ADP + hexose 6-phosphate

4. Phosphoglucomutase
   α-D-glucose-1,6-phosphomutase:α-D-glucose 1-phosphate   EC 5.4.2.2

   glucose 6-phosphate ↔ glucose 1-phosphate

5. Phosphoglucose isomerase
   D-glucose-6-phosphate ketolisomerase   EC 5.3.1.9

   glucose 6-phosphate ↔ fructose 6-phosphate

6. UDP-glucose pyrophosphorylase
   UTP:α-D-glucose-1-phosphate uridylyltransferase   EC 2.7.7.9

   UTP + glucose 1-phosphate ↔ UDP-glucose + PP_i

7. Sucrose phosphate synthase
   UDP-D-glucose:D-fructose 6-phosphate 2-α-D-glucosyltransferase   EC 2.4.1.14

   UDP-glucose + fructose 6-phosphate → sucrose 6-phosphate + UDP

8. Sucrose phosphatase
   Sucrose 6-phosphate phosphohydrolase   EC 3.1.3.24

   Sucrose 6-phosphate + H_2O → sucrose + P_i

9. Fructan exohydrolase (FEH)
   Exo-β-D-fructan fructohydrolase   EC 3.2.1.80

   $$\text{G-(F)}_n + H_2O \rightarrow \text{G-F-(F)}_{n-1} + F$$
   fructan polymer   fructose

10. Fructokinase
    UTP:D-fructose 6-phosphotransferase   EC 2.7.1.4

    UTP + fructose → UDP + fructose 6-phosphate

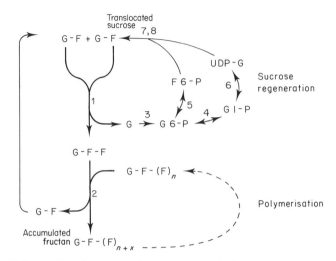

Fig. 4.3 Scheme for fructan polymer synthesis in the developing tubers of Jerusalem artichoke. The enzymes involved are indicated by the numbers given in Table 4.4.

The SST is able to use only sucrose, and not oligofructans of higher DP, as a fructosyl acceptor. Therefore SST cannot be responsible for the synthesis of polymeric fructans, and its role is limited to the synthesis of the trisaccharide. Polymerisation in fructan synthesis is catalysed by the second of the two principal enzymes responsible for fructan synthesis: fructan fructosyltransferase (FFT) (Table 4.4). This enzyme catalyses the reversible transfer of fructosyl residues between oligomeric and polymeric fructans. The direction of the reaction depends on the relative chain length and concentration of the available fructosyl donors and acceptors. Sucrose cannot act as a fructosyl donor for the reaction catalysed by FFT, but is a most effective fructosyl acceptor. The trisaccharide, 1-kestose, is an effective fructosyl donor, so that when it is incubated with oligomeric and polymeric fructans, FFT activity leads to chain elongation and the release of sucrose, which in turn becomes available for a SST-catalysed synthesis of 1-kestose. In this way, the concerted action of SST and FFT leads to the net synthesis of fructans from sucrose with the release of glucose. Tubers of Jerusalem artichoke are capable of synthesising sucrose from glucose (or fructose), and it is believed that the glucose released by the action of SST is simply recycled to sucrose by the conventional cytosolic pathway (Figure 4.3).

FRUCTAN DEPOLYMERISATION DURING STORAGE

When the supply of sucrose to the tubers runs dry in the late autumn, fructan synthesis ceases, SST activity disappears, and the tubers enter a period of relative dormancy. At this point about half of the fructan chains are polymeric, with DPs in the range 30–50 (Praznik & Beck, 1987). During the dormant period the fructans undergo a gradual depolymerisation: the mean DP decreases, the proportion of polymers with a DP >30 declines drastically, and there is a corresponding increase in the proportion of oligomeric chains and sucrose (Praznik & Beck, 1987). Depolymerisation is accelerated by low temperatures, and can be considered analogous to the cold-induced conversion of starch to sucrose seen in stored potatoes.

Depolymerisation is initiated by the hydrolytic release of the terminal fructosyl residue by the activity of a fructan exohydrolase (FEH) (Table 4.4). Activity of FEH in extracts of Jerusalem artichoke is specific for oligomeric and polymeric fructan, and it is virtually inactive against sucrose. The fructose liberated by FEH action is presumably converted to sucrose by the conventional pathway (Figure 4.4). The sucrose acts as fructosyl acceptor for the FFT-catalysed transfer of fructosyl units from polymeric fructans. Thus the concerted action of the FEH and FFT, together with the ancillary enzymes of sucrose synthesis, combine to reduce the mean DP of the stored fructan (Figure 4.4), so that there is no net loss of fructan during depolymerisation, merely a reshuffling of fructosyl residues from the longer chains to a larger number of shorter chains.

When the tubers sprout in the spring, the fructan, by now predominantly in the oligomeric form, is mobilised by FEH action, and the released fructose is converted to sucrose for export to the growing parts of the plant.

The three stages of fructan metabolism: synthesis, depolymerisation and mobilisation are governed by two kinds of factors. First, a change in enzyme complement, with synthesis depending on high SST activity, and both

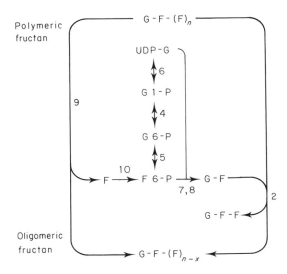

Fig. 4.4   Scheme for the depolymerisation of fructan during the storage of Jerusalem artichoke tubers. The enzymes involved are indicated by the numbers given in Table 4.4.

depolymerisation and mobilisation depending on FEH activity. Activity of the FFT remains high during all three stages. The second governing factor is the altered relationship between the stored fructan and external sucrose. During synthesis there is an abundant supply of sucrose which effectively drives the system towards polymerisation. During tuber dormancy, especially at low temperatures, sucrose is generated from the fructose released by FEH action, but this sucrose has no external sink, and is therefore available as a fructosyl acceptor by the action of FFT. During mobilisation the sucrose generated from fructan depolymerisation is drawn away from the pool of stored fructan by the growing regions of the developing plant. Thus the internally regulated enzyme complement of the tubers, together with the externally regulated supply and demand of sucrose, act to control fructan polymerisation and depolymerisation in the Jerusalem artichoke tuber.

SUBCELLULAR LOCALISATION OF FRUCTAN METABOLISM

As soon as the principal enzymes responsible for fructan synthesis and breakdown had been identified, a proposal was made by Edelman and Jefford (1968) for their subcellular localisation. It was suggested that SST was confined to the cytosol, where it was well supplied with incoming sucrose to generate 1-kestose, which acted as a fructosyl donor to the FFT, located at the tonoplast so that fructosyl units were accepted from the cytosolic face and donated to the elongating fructans at the internal face of the tonoplast. In this way FFT catalysed a transmembrane conduction of fructosyl units from a sucrose-rich cytosol to a relatively sucrose-poor vacuolar sap in which the fructans accumulated. This was an elegant solution to the problem which was perceived to arise from the observation that sucrose, acting as a preferential acceptor of fructosyl units in *in vitro* FFT-catalysed transfructosylation reactions, effectively inhibited the generation of high DP polymers. The model of fructan synthesis suggested by Edelman & Jefford (1968) remained untested for nearly 20 years, until it was shown that vacuoles isolated from Jerusalem artichoke tubers contain both SST and FFT activities (Frehner *et al.*, 1984), and that both enzymes were in the vacuolar sap (Darwen & John, 1989). However, more work is needed before we can be really sure that we have identified the site of fructan synthesis in the storage parenchyma cells.

BIOTECHNOLOGICAL DEVELOPMENT

The two species generally considered as sources for large-scale fructan production are chicory and Jerusalem artichoke. Both plants can be readily cultivated and harvested, and their incorporation into current crop rotation schemes would provide a beneficial widening of the crop range. Chicory has the advantage that it is already grown in Belgium and The Netherlands as a salad crop, and its agronomic requirements resemble those of sugar beet

(Chubey & Dorrell, 1978). The Jerusalem artichoke can reach a height of 4 m, and a relatively large fraction of the total biomass goes to form the extensive stem structure. By contrast, chicory forms only a rosette of leaves above its swollen root, so that a larger proportion of the total assimilate can be allocated to fructan accumulation. Thus chicory is likely to be the crop of choice for the production of fructan. However, Jerusalem artichoke yields well over a range of climates within temperate regions, and it is well adapted to a wider variety of soils than is chicory. Consequently Jerusalem artichoke may become important in less favoured regions. Techniques for producing high-fructose syrup from Jerusalem artichoke tubers have already been described (Kierstan, 1980). Figure 4.5 gives an outline of such a process.

The main technical problem to be overcome for the commercial production of fructan as a major crop product is related to the storage of the chicory roots and Jerusalem artichoke tubers. In order to utilise a processing factory economically, it is necessary to run the factory for as long a period as possible. This requires a continuous supply of raw material of a consistently high quality. Storage at low temperatures would help maintain tuber quality, but this would accelerate what has been called 'the crucial factor' (Fuchs, 1987) in the economic feasibility of Jerusalem artichoke as an industrial crop: the depolymerisation that occurs on storage at low temperatures (Praznik & Beck, 1987). The production of high-fructose syrups and furan-based chemicals demands a high DP in the fructan, not only to maintain a high fructan: glucose ratio, but also to facilitate purification. Depolymerisation is less of a

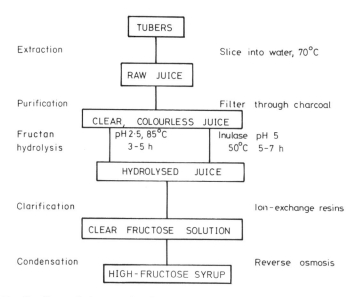

Fig. 4.5 Outline of the production of a high-fructose syrup from Jerusalem artichoke tubers (modified from Fleming & GrootWassink, 1979).

problem where the tubers are destined for fermentation or feed (Table 4.1). A variety of technical solutions to prevent depolymerisation have been sought (Kosaric *et al.*, 1984), but the most cost-effective and permanent means of maintaining a high DP during storage would undoubtedly be to obtain varieties in which expression of FEH is delayed so that depolymerisation occurs only on regrowth, and not when the tubers are stored.

FURTHER READING

Edelman, J. & Jefford, T.G. (1968). The mechanism of fructosan metabolism in higher plants as exemplified in *Helianthus tuberosus, New Phytologist* **67**, 517–531.

Fleming, S.E. & GrootWassink, J.W.D. (1979). Preparation of high-fructose syrup from the tubers of the Jerusalem artichoke (*Helianthus tuberosus* L.), *CRC Crit. Rev. Food Sci. Nutr.* **12**, 1–28.

Fuchs, A. (1987). Potentials for non-food utilization of fructose and inulin, *Starch/ Stärke* **39**, 335–343.

John, P. (1991). Fructan quality and fructan synthesis, *Biochem. Soc. Trans.* **19**, 569–572.

Kosaric, N., Cosentino, G.P., Wieczorek, A. & Duvnjak, Z. (1984). The Jerusalem artichoke as an agricultural crop, *Biomass* **5**, 1–36.

Pollock, C.J. (1984). Physiology and metabolism of sucrosyl-fructans, in *Storage Carbohydrates in Vascular Plants*, Ed Dennis, D.H., Cambridge, Cambridge University Press, pp. 97–113.

Pollock, C.J. & Chatterton, N.J. (1988). Fructans, in *The Biochemistry of Plants. Volume 14 Carbohydrate*, Ed Preiss, J., New York, Academic Press, pp. 109–140.

Pontis, H.G. & Del Campillo, E. (1985). Fructans, in *Biochemistry of Storage Carbohydrates in Green Plants*, Eds Dey, P.M. & Dixon, R.A. New York, Academic Press, pp. 205–227.

ADDITIONAL REFERENCES

Anon. (1987). Crystalline fructose: a breakthrough in corn sweetener process technology, *Food Technol.* **41**, 66–72.

Chubey, B.B. & Dorell, D.G. (1978). Total reducing sugar, fructose and glucose concentrations and root yield of two chicory cultivars as affected by irrigation, fertilizer and harvest dates, *Can. J. Plant Sci.* **58**, 789–793.

Darwen, C.W.E. & John, P. (1989). Localization of the enzymes of fructan metabolism in vacuoles isolated by a mechanical method from tubers of Jerusalem artichoke (*Helianthus tuberosus* L.), *Plant Physiol.* **89**, 658–663.

Frehner, M., Keller, F. & Wiemken, A. (1984). Localization of fructan metabolism in the vacuoles isolated from protoplasts of Jerusalem artichoke tubers (*Helianthus tuberosus* L.), *J. Plant Physiol.* **116**, 197–208.

Hendry, G. (1987). The ecological significance of fructan in a contemporary flora, *The New Phytologist* **106** (Suppl.), 201–216.

Incoll, L.D. & Neales, T.F. (1970). The stem as a temporary sink before tuberization in *Helianthus tuberosus* L., *J. Exp. Bot.* **21**, 469–474.

Kierstan, M. (1980). Production of fructose syrups from inulin, *Process Biochem.* **15**, 2–4, 32.

Phelps, C.F. (1965). The physical properties of inulin solutions, *Biochem. J.* **95**, 41–47.

Praznik, W. & Beck, R.H.F. (1987). Inulin composition during growth of tubers of *Helianthus tuberosus*, *Agric. Biol. Chem.* **51**, 1593–1599.

Chapter 5

Cellulose

INTRODUCTION

Cellulose is the component of plant cell walls that provides the high tensile strength which enables plants to grow in the terrestrial environment. As a crop product, cellulose is most easily recognised in the form of cotton fibres (mainly from the cotton plant), which are almost pure cellulose, and as wood, which takes its strength from the 40 to 55 per cent cellulose content. Cellulose also provides the basis of a range of minor natural fibres, such as jute and sisal. Enormous quantities of cellulose are used to manufacture paper, cardboard, and related materials. In addition to being used in its natural form, cellulose is the raw material for the manufacture of acetate films, rayon, and other man-made fibres (Table 5.1).

Cellulose is made up of linear, flat, ribbon-like chains of β-1,4-linked D-glucose residues (Figure 5.1). The degree of polymerisation (DP) of the bulk of the cellulose of cotton and wood fibres is estimated to be about 14 000.

The fibres of commerce consist of the cell walls of specialised cells. In the case of cotton these cells take the form of long, unicellular hairs which grow out from the outer layers of the seed coat, eventually attaining a length of about 25 mm and a diameter of about $20\,\mu$m (Figure 5.2). When the fibre is mature, the protoplast of the fibre cell dies, the cell wall flattens and becomes convoluted. This conformation helps the individual fibres to adhere to one another when the fibres are twisted together during the production of yarn by spinning. Shorter fibres are also formed by the seed coat. These are called lintners and are used as a source of industrial cellulose.

The other cellulose-based fibres are far less important than cotton in terms of production levels (Figure 5.3), and structually they are also quite different from cotton. For example, the fibres from flax, which are woven into linen,

Table 5.1 Products made from cellulose.

Starting material	Products
Natural fibres	Cotton textiles Papers Linen Fillers in plastics
Viscose	Rayon fibres Cellophane film
Esters	Acetate fibres Acetate films Nitrate explosives Nitrate varnish
Ethers	Carboxymethylcellulose Hydroxyethylcellulose

consist of bundles of thick-walled fibre cells which run the length of the stem in a ring immediately outside the phloem. The cell walls thicken as the cellulose is deposited, progressively from the outer to the inner fibre cells.

The fibres of wood are made up largely of the tracheids of the secondary xylem. They are about the same width as cotton fibres but are shorter, extending only up to 5 mm in length.

The principal source of cellulose for paper is wood pulp, but when locally available, other sources become important, often as crop residues, such as cereal straw, cotton lintners, and bagasse from sugar cane. The fibres in the raw materials are separated by treatment with alkali or sulphite, which partly removes the lignin and other non-cellulosic components. When the fibres are removed from aqueous suspension they are pressed and dried. This procedure results in the fibres bonding together to form a layered network. The inter-fibre bonding is important in determining the strength of the paper.

Viscose is the name given to an alkaline solution of cellulose obtained by dissolving cellulose in strong alkali and then treating it with carbon disulphide to form cellulose xanthate. When the solution is filtered and acidified the

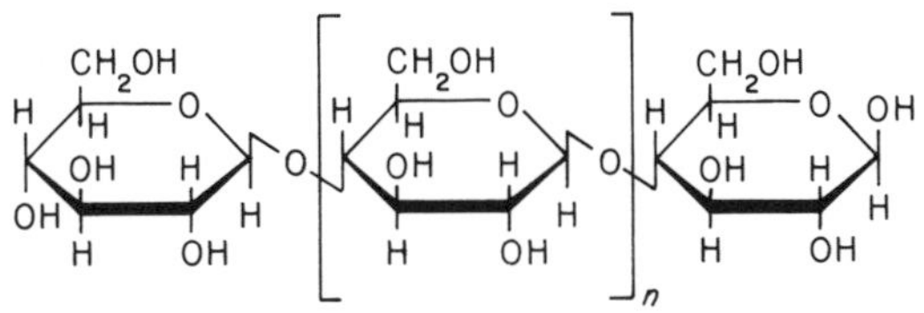

Fig. 5.1 The chemical structure of cellulose.

Fig. 5.2   Development of the cotton fibre. A, ovule, 1 day after anthesis showing fibre initials (× 36). B, ovule, 2 days after anthesis; the arrow indicates where fibre initials first appear (× 36). C, fibres with blunt tips, early in third day after anthesis (× 72). D, fibres with tapered tips, late in third day after anthesis (× 59). (From Stewart (1975), by kind permission of the author and publisher.)

cellulose is regenerated. During this regeneration the cellulose can be made into a thin film (as in cellophane) or into silk-like textile fibres (rayon) which can be spun into strong artificial fibres. Cellulose is also converted to the acetate and nitrate esters on a large scale; the resulting derivatives possess properties which make them suitable for particular applications. Thus cellulose acetate is used for textiles, cigarette filters and photographic film; while cellulose nitrate, once used for photographic films, is now used as an explosive, as guncotton, and in varnishes and other coatings. The cellulose ethers are water-soluble derivatives with innumerable uses in industry: carboxymethylcellulose is frequently used to thicken aqueous solutions in the food

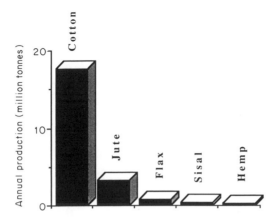

Fig. 5.3   Annual world production of cellulose-based fibres (averages for years 1987–89, from FAO Production Yearbook, 1989).

industry, in paints, printing inks, cosmetics, and in the oil industry; hydroxyethylcellulose is incorporated in a range of products from cements to toothpastes.

STRUCTURE OF CELLULOSE FIBRES

Both the chemical and physical properties of cellulose are determined to a large extent by the aggregation of the β-1,4 chains into structural fibrils (Figure 5.4). Each fibril contains from thirty to several hundred polymeric chains arranged in an ordered crystalline structure in which the flattened β-1,4-glucan chains run parallel to each other, with the hydroxyl groups exposed at the sides. These hydroxyl groups take part in hydrogen bonding, with linkages both within the polmeric molecules and between them (Figure 5.4). This arrangement of the hydroxyl groups in cellulose makes them relatively unavailable to a solvent such as water, in comparison to the hydroxyl groups of amylose (Chapter 3), where the glucan chains are α-rather than β-linked. Thus the relative insolubility of cellulose can be understood as a stereochemical consequence of its 1,4-linkages being of a β- and not an α-conformation.

The intermolecular hydrogen bonds hold together the flat cellulose chains laterally so that molecular sheets are formed. The surfaces of the polymeric chains are covered by hydrogen atoms linked directly to carbon, and these provide the basis of a hydrophobic attraction between the sheets of chains. It is this combination of the intra- and intermolecular hydrogen bonding with the hydrophobic attraction within the fibril that gives cellulose its unusual resistance to chemical attack, as well as its high tensile strength.

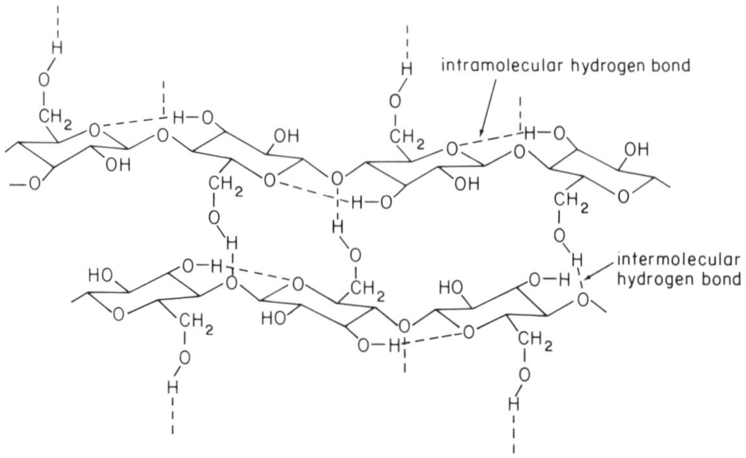

Fig. 5.4 Hydrogen bonding between the β-1,4-glucan chains in a microfibril of cellulose I (redrawn from Alberts *et al.*, 1983, with permission).

The native celluloses from a variety of plant sources have a similar crystallographic structure, which is referred to as cellulose I. If, however, cellulose is dissolved in an appropriate solvent and precipitated, or if it is treated with an alkaline swelling agent and then washed with water, the cellulose crystallises into a different form, referred to as cellulose II. The major difference between the two forms is that adjacent chains in cellulose II run antiparallel, that is, with the reducing ends at opposites ends of the chain. It is not entirely clear how this reversal is brought about. One possibility is that dissolving the cellulose provides an opportunity for chains from adjacent microfibrils to intermix, and, while individual chains within a microfibril of native cellulose run parallel to each other, adjacent microfibrils run antiparallel (French, 1985). The conversion of cellulose I to cellulose II by treatment with alkali is the basis of the process termed mercerisation, which has been used for many years to improve the dye affinity, tensile strength and smoothness of cotton fabrics. Although the major part of the cellulose is organised in a crystalline state, there are amorphous, disordered zones along the microfibrils. The physico-chemical basis of these zones is unclear.

The arrangement of the microfibrils in the layers of the cotton fibre is shown in Figure 5.5. The outside of the cotton fibre is covered by a waxy cuticle, which immediately surrounds the thin primary cell wall. In this wall the cellulose fibrils are arranged in a randomly interlaced network. Internally lies the secondary wall, which consists of three layers. In the outer of these layers, the S_1 layer, the microfibrils are arranged in a banded pattern of alternate close and open packing; in the middle, S_2 layer, and inner, S_3 layer, the microfibrils spiral at an angle with respect to the fibre axis. The bulk of the cellulose of the cotton fibre is represented by the S_2 layer. The central lumen

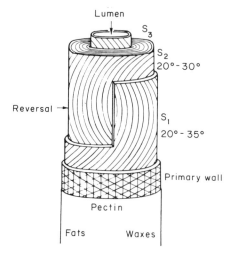

Fig. 5.5 Morphological structure of the cotton fibre (from Young, 1986, with permission).

represents the space taken up by the plant protoplasm, only vestiges of which remain in the mature fibre.

The walls of tracheids and xylem vessels lack the cuticle of the cotton fibre, and are surrounded *in situ* by the middle lamella (Figure 5.6). But in other respects the walls of wood fibres resemble those of cotton fibres. As in the cotton fibre, the primary cell wall of wood fibres contains cellulose micro-

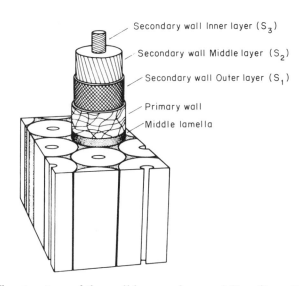

Fig. 5.6 The structure of the wall layers of a wood fibre (from Young, 1986, with permission).

Table 5.2 The composition of cotton and wood fibres (data from Bacic *et al.*, 1988; Salme'n, 1985).

	Cotton fibre	Wood
Cellulose	94	40–55
Non-cellulose	5†	35–50‡
polysaccharides		
Lignin	0	15–36
Wax	<1	0

†Contains 0.4–0.7 per cent callose.
‡In softwoods the major hemicellulose is *O*-acetyl-galactogluco-mannan with lesser amounts of arabino-(4-*O*-methylglucurono)xylan. In hardwoods *O*-acetyl-(4-*O*-methylglucurono)xylan dominates with lesser amounts of glucomannan.

fibrils which are randomly arranged in an interlaced network, and the secondary wall consists of three layers, the thickest being the middle, S_2 layer. In the outer, (S_1) and inner (S_3) layers the microfibrils run in alternating helical patterns at a large angle to the fibre axis.

The non-cellulosic components of the cell wall are pectin, protein, lignin and hemicelluloses (Table 5.2). Pectins are acidic polysaccharides consisting of blocks of galacturonic acid and methylgalacturonic acid residues interspersed with rhamnose residues to which side chains of neutral oligo and polysaccharides (arabinans, galactans and arabinogalactans) are attached. They are more important in growing than in non-growing walls, and consequently they are an insignificant constituent of the cellulose fibres of commerce.

Some of the protein in the cell wall is enzymic, but much of it is structural. The best characterised of the structural proteins are the glycoproteins called extensins, which are rich in the unusual amino acid hydroxyproline, which is represented in numerous repeat sequences of a pentapeptide (ser-hyp-hyp-hyp-hyp) of serine (ser) and hydroxyproline (hyp).

Lignin is a complex, aromatic polymer with a molecular weight of about 11 000 formed by the three-dimensional polymerisation of phenylpropane derivatives. Lignification renders the wall less permeable to water and more rigid, and consequently it is a characteristic of the secondary walls of cells with water-conducting and supporting functions. The walls of mature tracheids contain 15 to 36 per cent lignin.

The hemicelluloses consist of a heterogenous group of branched polysaccharides. The particular constitution of the hemicellulose polymer depends on the particular species and on the tissue, but glucose, xylose and mannose often predominate. The composition and general structure of the xyloglucan hemicellulose of cotton fibres is shown in Figure 5.7.

These non-cellulosic components do not take the form of ordered structures like the cellulose microfibrils. It has therefore been customary to see the

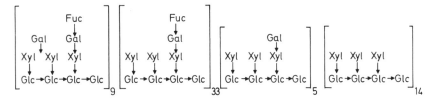

Fig. 5.7 Oligomeric subunits of the xyloglucan extracted from cotton fibres. The numbers indicate the proportions in which the subunits are represented in the polymer. (Redrawn from Hayashi & Delmer, *Carbohydrate Research*, **181**, 273–277, 1988. Reproduced by kind of permission of the authors and publishers.)

non-cellulosic components as forming an amorphous matrix within which the cellulose microfibrils are embedded. However, it will become apparent when we consider the biosynthesis of cellulose that the hemicelluloses at least may well have a more dynamic role in directing the pattern of cellulose deposition than is implied by their presence as an amorphous matrix. Even in the cotton fibre, where the cellulose of the secondary wall is laid down in almost pure form, its deposition probably depends on an interaction with the other wall constituents. An understanding of cellulose biosynthesis requires therefore that we know something of the relationship between cellulose and the non-cellulosic cell wall components.

A useful starting point in examining the wall architecture is to note that, except for cellulose, the wall polymers are intrinsically soluble, but neverthe-less the plant cell wall matrix cannot easily be extracted with water. The implication of this observation is that the constituent polymers are cross-linked (Fry, 1988).

What kinds of cross-linking integrates the cellulose microfibrils with the other components? Evidence comes largely from analysis of the fragments of the wall polymers released after specific hydrolysis with purified glycanases. The evidence suggests that the rhamnogalacturonan pectin, xyloglucan hemi-celluloses and extensin are all connected by covalent bonds, whereas hydro-gen bonds interconnect xyloglucan and glucoxylan hemicelluloses to the cellulose microfibrils (Hayashi, 1990). Thus the cellulose microfibrils are attached via hydrogen bonds to a macromolecular complex of other cell wall components. This implies that the hemicelluloses have an important role to play in the integration of cellulose in the cell wall, both in a dynamic sense during cell wall synthesis, and also in stabilising the mature cell wall.

In the cotton fibre the main hemicellulose, xyloglucan, is deposited only during the elongation stage. When extracted from the developing fibre the xyloglucan is found to have a molecular weight of 80 000, and to consist of at least four different oligosaccharide subunits (Hayashi and Delmer, 1988). The structure of these four subunits is shown in Figure 5.7. The linear sequence of these subunits in the xyloglucan is not yet known, but the general pattern consists of a glucan backbone from which extend short side-chains, predominantly of xylose.

In hardwood the main hemicellulose components are 4-*O*-methylglucuronoxylans. In the wood of the Linden tree they have been shown (Vian *et al.*, 1986) to be concentrated in the transition zone between the S_1 and S_2 layers of the secondary wall. Since the direction in which the cellulose fibres rotate around the fibre differs between the two successive layers, this transition zone marks a region where the cellulose fibres change direction abruptly during development. Thus it has been proposed (Vian *et al.*, 1986) that the glucuronoxylans have a morphogenic role by giving direction to the assembly of the microfibrils of cellulose.

In the cell wall the xyloglucan is likely to be bound to cellulose so that the β-1,4-glucan backbone of the xylan lines up with the β-1,4-glucan chains of the cellulose, thus facilitating the hydrogen bonding between the two components, with the xylan side-chains having an interconnecting function. In the primary cell wall xyloglucan probably interweaves into the non-crystalline parts of the microfibrils.

INCORPORATION OF CELLULOSE INTO FIBRE CELL WALLS

Cell walls are formed *de novo* when a plant cell divides. Before the primary cell wall is laid down, its position in the dividing cell is accurately marked by a circumferential band of microtubules which forms just below the plasma membrane, defining the plane in which the new cell wall will divide the parent cell. After the newly separated chromosomes have moved towards the cell poles, membrane vesicles, directed by microtubules, aggregate to form the cell plate. These membrane vesicles originate in the Golgi apparatus and contain pectic polysaccharides and hemicelluloses. When the vesicles coalesce they form the new plasma membranes of the adjacent daughter cells, and their contents form the middle lamella of the new wall. It is after these vesicles have coalesced that cellulose microfibrils begin to appear. As the primary cell wall is formed the microfibrils are deposited, usually in an orientation transverse to the longitudinal axis of the cell. In the primary cell wall cellulose is, however, only a minor component. A large part of the primary wall is made up of matrix polysaccharides and protein. These components are delivered to the growing primary wall by the fusion of membrane vesicles derived from the Golgi apparatus.

The loosening of bonds in the multimolecular complex of the primary cell wall that accompanies cell expansion is believed to be associated with an acidification of the cell wall. The evidence can be summarised as follows. First, auxins, which stimulate growth, also stimulate the secretion of acid into the wall solution. Second, the growth-stimulating effect of auxins can be mimicked by exogenous acid. Third, auxin-stimulated growth can be blocked by buffering the wall solution at a more alkaline pH. The identification of the bonds broken by acidification has not proved to be easy, but it seems likely that they are in the xyloglucan matrix rather than in the cellulose microfibrils.

The growth of the commercially important fibres of cotton and wood is strongly directional; the length of the mature cell is hundreds or thousands of times greater than the width. The driving force for cell expansion, the turgor pressure, is applied equally in all directions, and the ability of plant cells to expand unindirectionally is due in large part of the unequal resistance to growth offered by transverse and longitudinal cell walls because of the different orientation of the cellulose microfibrils in these walls. The microfibrils are virtually unstretchable, and therefore for the cell to enlarge the fibrils must slide past each other. In elongating cells, such as the hairs of cotton seeds and the xylem tracheids, the microfibrils of the primary wall are laid down perpendicular to, rather than parallel to, the longitudinal axis, so that in effect they form numerous hoops surrounding the cylindrical cell. This particular organisation of the primary cell wall makes the cell much more extensible in the longitudinal direction than in the tangential. However, as the cell grows the initial orientation becomes lost. As each layer is pushed outward in the wall it becomes increasingly strained, so that there is a gradual shifting of the microfibrillar arrangement. The original transverse arrangement becomes random and finally it is predominantly longitudinal (Figures 5.5, 5.6).

The secondary wall of cotton seed hairs consists essentially of $5-10\,\mu m$ layer of almost pure cellulose applied as a sequence of lamellae (Figure 5.5). Deposition of the secondary cell wall begins about 18–28 days after flowering and takes about 24–30 days to complete. Although it is clearly a secondary wall, elongation growth of the cotton fibre continues to some extent during its deposition. In contrast to the arrangement of microfibrils in the primary wall, the microfibrils of the secondary wall are arranged as co-axial helices with a 20° to 30° angle to the longitudinal axis. They are not straight but take the form of undulating waves (Figure 5.5).

Because cotton fibres elongate synchronously over a period of about 3 weeks it is not difficult to acquire sufficient material at each stage to analyse biochemically the changes that accompany the development of the cotton fibre (Meinert & Delmer, 1977). Figure 5.8 shows the pattern of cellulose accumulation discovered from these experiments. The initiation of secondary wall formation coincides with a relative increase in the cellulose deposited, and a sharp decrease in the content of pectin and protein, so that at maturity cellulose accounts for over 90 per cent of the fibre. The composition of the primary cell wall of the cotton fibre resembles that of the primary cell wall of other plant tissues, but with the development of the secondary wall and the enormous growth of the fibre, the relative amounts of the individual neutral sugars change. The contents of xylose and non-cellulosic glucose increase, while that of arabinose decreases. Xylose and non-cellulosic glucose are the main sugars which remain associated with cellulose after it has been extensively washed to free it of non-cellulosic materials. Thus these data from cotton indicate a close structural relationship between the cellulose and a

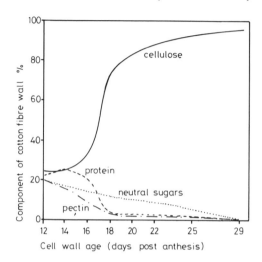

Fig. 5.8   Composition of the cell wall during the development of the cotton fibre (data from Meinert & Delmer, 1977).

xyloglucan hemicellulose. That relationship will be recalled when we turn to examine the difficult subject of the actual biochemistry of cellulose biosynthesis.

BIOSYNTHESIS OF CELLULOSE

There is a unique problem which introduces an element of uncertainty into almost any statement that is made about the biochemical basis of cellulose synthesis: enzyme systems capable of synthesising authentic cellulose I *in vitro* have not been prepared from higher plants. The cellulose synthase is sensitive to cell disruption so that activity disappears completely when cell integrity is lost. Investigators who in the past have claimed to have shown cellulose synthesis in cell-free homogenates have been misled by the incorporation of labelled glucosyl residues from what are believed to be the precursors of cellulose into other polymers, which invariably have included the β-1,3-glucan, callose.

In the developing fibres of cotton seed UDP-glucose is the most abundant nucleoside diphosphate glucose, and it is favoured by current evidence as the substrate for cellulose synthesis. Nevertheless until an authentic cellulose synthase activity has been demonstrated *in vitro* the identity of the donor remains uncertain. Small amounts of β-1,4-glucan can be synthesised *in vitro* with membrane preparations from some plants, if a low concentration ($<50\,\mu$M) of UDP-glucose is used and Mg^{2+} is added. However, this is unlikely to be even a limited activity of the cellulose synthase, since the

Cellulose

activity is not specifically associated with plasma membranes, and it can be accounted for by the β-1,4-glucan synthase that is located in the Golgi and is responsible for the glucan backbone of xyloglucan.

The ultimate source of carbohydrate for cellulose synthesis in fibres is sucrose, which can yield UDP-glucose most readily by the action of sucrose synthase (see Chapter 2), an enzyme which is widely distributed in cellulose synthesising tissues:

$$\text{sucrose} + \text{UTP} \leftrightarrow \text{UDP-glucose} + \text{fructose 6-phosphate}$$

The UDP-glucose could also be generated from hexose-phosphates by the action of the appropriate pyrophosphorylase:

$$\text{glucose 6-phosphate} + \text{UTP} \leftrightarrow \text{UDP-glucose} + \text{PP}_i$$

The pectic and hemicellulose polymers and callose are all synthesised from UDP and GDP derivatives of the appropriate monosaccharides. UDP-glucose is converted to the UDP derivative of galacturonic acid, glucuronic acid, xylose, arabinose, and rhamnose by epimerase, oxido-reductase and carboxylase reactions (Figure 5.9). Mannose is donated as the GDP derivative which is formed directly from mannose 1-phosphate and GTP by a pyrophosphorylase. This pool of nucleoside diphosphate sugars is drawn upon by the synthases for the stereospecific glycosyl transfer involved in polymerisation (Figure 5.9).

In the case of the pectic and hemicellulosic polymers the synthases are bound to the membranes of the Golgi and endoplasmic reticulum. The Golgi apparatus is the location of the enzyme responsible for the transfer of glucosyl residues to the growing chain of the hemicellulose, xyloglucan. This enzyme is known as glucan synthase I to distinguish it from the glucan synthase II responsible for glucosyl transfer to the growing cellulose chain, which is believed to be located at the plasma membrane.

It is known from work using radioactively labelled precursors that pectin and hemicelluloses are transported to the plasma membrane within small vesicles which bud off from the Golgi apparatus, and discharge their contents to the developing cell wall as the vesicles fuse with the plasma membrane (Figure 5.9).

Unlike the non-cellulosic polysaccharides, cellulose is never found within the plant cell. Therefore the polymerisation of the cytoplasmically located precursors involves translation of glucosyl residues across the plasma membrane to the innermost face of the cell wall where it is laid down as the crystalline cellulose I. It has long been widely accepted that the synthase is an integral protein of the plasma membrane. In particular it appears to be associated with a particular structural feature of the plasma membrane: viz., the rosette of granules which is observed when the plasma membrane is split open within the plane of the membrane by the technique of freeze-fracture electron-microscopy (Figure 5.10).

The putative cellulose synthase has also been studied in algae where the microfibrils are larger than those of higher plants and with a higher crystallinity.

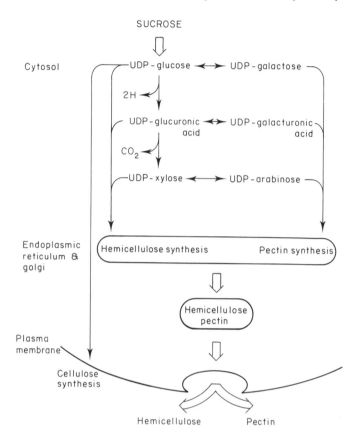

Fig. 5.9 Pathways followed in the biosynthesis of cell wall polysaccharides (based on Northcote, 1984).

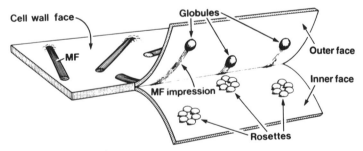

Fig. 5.10 Stylised drawing of terminal complexes observed by freeze-fracture of the plasma membrane of higher plant cells. MF, microfibril. (From Delmer, 1987. Reproduced by kind permission of the author and publishers from the *Annual Review of Plant Physiology* Vol. 38, © 1987 by Annual Reviews Inc.)

In these algae, freeze-fracture of the plasma membrane reveals a linear series of particles, termed linear complexes, which probably traverse the membrane. In higher plants the membrane particles are organised into hexagonal arrays of rosettes, which are seen at the inner face of the plasma membrane after it has been freeze-fractured (Figure 5.10). On the outer face of the membrane, globules can be seen at sites corresponding to the positions of the rosettes (Figure 5.10). This has led to the theory that the globules sit within the rosettes. The central piece of evidence that identifies these arrays of rosettes and globules with the cellulose synthase is that they are seen at the ends of microfibrils. In addition the rosettes show a lability which parallels the lability of the catalytic activity as a cellulose synthase. Thus it has been found that the rosettes are observable only if the plant material is handled with great care. However, the actual role of these structures in cellulose biosynthesis will become clear only when their enzymatic activity has been demonstrated *in vitro*.

It is assumed that the force exerted by the polymerisation of glucosyl units into cellulose pushes the rosette structures along the plane of the membrane. Support for this idea comes from studies of a bacterium, *Acetobacter xylinum*, which synthesises cellulose. It makes cellulose not as a component of its cell wall, but instead as a single large extracellular ribbon. A row of synthetic sites along one of the long axes of the cell produce the microfibrils which associate to form the ribbon. The force generated during the synthesis of this ribbon drives the bacterial cell backwards in the medium.

The direction in which the plant rosette structures move through the plasma membrane will determine the orientation of the cellulose microfibril in the cell wall. The electron microscope reveals that the most recently made microtubules are laid down in a pattern that reflects the pattern taken by microtubules below the plasma membrane. Consistent with this directing role for the microtubules, it has been found that agents, such as colchicine, which disrupt microtubules, also disorientate the newly synthesised microfibrils. However, what is not yet explained is how the pattern of the microtubules is transmitted to the synthase complexes. A possible mechanism by which the microtubules could determine microfibril orientation is by directing the movement of the secretary vesicles which bring xyloglucan to be secreted at the plasma membrane. It was described earlier in this chapter how the cellulose microfibrils appear to be integrated into the cell wall structure via xyloglucan to which the microfibrils are linked by hydrogen bonds, and how auxin-induced acidification of the cell wall loosened linkages in the xyloglucan. It will also be recalled that in cotton fibres the difficulty of freeing cellulose of xylose and non-cellulosic glucose spoke for a close structural relationship between the cellulose and xyloglucan. A role for xyloglucan in integrating the nascent cellulose into the cell wall would therefore only add a dynamic element to the already established importance of xyloglucan in the organisation of cellulose in the cell wall.

Little can be said about the mechanisms involved in the initiation or termination of microfibril synthesis. However, there is evidence that a precise control over the chain length is applied in cotton as in other plants. Marx-Figini (1969) found that the DP of the cotton cellulose showed a bimodal distribution, with cellulose in the primary wall having DPs between 2000 and 6000, while in the secondary wall the DP was around 14 000. This value was not affected by the rate of synthesis, which was varied experimentally by altering the conditions under which the plants grew, and the DPs were independent of the stage in the development of the secondary wall. From these observations Marx-Figini (1969) concluded that:

> the biosynthesis of secondary wall cellulose in higher plants. . . must proceed by a mechanism completely different from any other polymerisation process known in polymer chemistry. . . . Only a structure-controlled mechanism (for example a template mechanism) and not a time-controlled one can be considered, i.e. the degree of polymerisation of the secondary wall cellulose must be genetically determined.

Glucan synthase II

When a cell active in cellulose synthesis is damaged, it stops making cellulose and makes instead the β-1,3-glucan, callose. Callose is not normally a constituent of the cell walls of mature plant cells, being only transiently deposited at the onset of secondary thickening, but it is produced in large amounts in response to chemical or physical wounding. The enzyme responsible for callose synthesis, glucan synthase II, uses UDP-glucose as substrate and is located at the plasma membrane. These circumstances have led to the suggestion (Delmer, 1987) that the glucan synthase II enzyme is the same enzyme as the cellulose synthase, but has 'gone awry' when it synthesises callose *in vitro*. It is suggested that cell disruption alters the catalytic properties of the enzyme so that the polymeric product is changed from the highly ordered microfibrils of a β-1,4-glucan to the non-crystalline β-1,3-glucan. This sensitivity of the catalytic properties of the cellulose synthase is in line with the lability of its putative structural organisation in the membrane, where the rosette arrangement of particles is known to be unstable and disappear if the tissue is mistreated.

Three additional sets of observations are relevant to this enzyme duality theory.

First, the herbicide 2,6-dichlorobenzonitrile (DCB) is a specific inhibitor of cellulose synthesis when applied at μM concentrations. It does not inhibit the large amounts of β-1,3 glucan and small amounts of β-1,4-glucan that are made *in vitro* by plant preparations, nor does it inhibit the synthesis *in vivo* of the other cell wall polysaccharides. This specificity suggests that its mode of action is via an interaction with a vital component of the cellulose synthesising system. That component has been identified in cotton fibres as a polypeptide of 18 kDa. When the subcellular location of the polypeptide was determined it was, surprisingly, not found to be associated with the plasma membrane, but largely with the soluble phase (Delmer, 1987). If it is assumed that during cellulose synthesis *in vivo* the essential 18 kDa polypeptide is associated with the plasma membrane, then disruption of the cell could be responsible for its dissociation from the membrane.

Second, β-1,3-synthase by the glucan synthase II *in vitro* is activated by Ca^{2+}, which is normally present in the cell wall solution in mM concentrations but in the cytosol in μM concentrations. Thus, activity of the glucan synthase II in the cytosol would normally be restricted by the relatively low Ca^{2+} levels. An increased plasma membrane permeability to Ca^{2+}, or a loss of cell integrity would make the Ca^{2+} of the cell wall available to the glucan synthases II for β-1,3 synthesis.

Third, in *Acetobacter xylinum* and in membrane vesicles prepared from the plasma membrane of cotton fibres, there is evidence that a membrane potential is required for β-1,4-glucan synthesis. The relationship of the membrane potential to glucosyl transfer is quite unknown, but it does provide us with another explanation for the apparent dependence of cellulose synthesis on an intact plasma membrane, since the loss of membrane integrity would inevitably mean that the membrane potential would disappear.

These sets of observations form the experimental basis for a model of cellulose synthesis that incorporates a dual role for the glucan synthase II. In this model, cellulose synthesis can proceed in the intact cell catalysed by the glucan synthase II complexed with the 18 kDa polypeptide, and with the plasma membrane maintaining a potential which is negative internally and a low cytosolic levels of Ca^{2+}. Binding of DCB would result in the 18 kDa polypeptide becoming dissociated from the synthase so that cellulose synthase ceased. When the cell became damaged the membrane potential would disappear, Ca^2 would flow in, the 18 kDa polypeptide would again become dissociated, and cellulose synthesis would again cease. Now, however, the catalytic activity of the synthase would be directed to the synthesis of callose.

An attractive feature of this model for the plant glucan synthase II is that it explains how plant cells have available, in the form of the cellulose synthesising system, a large reserve capacity for callose synthesis that could be drawn upon without delay when required to contain damage to a plant tissue: the swift arming of a merchant marine to form a defensive navy. However, it must be emphasised that the model requires much experimental testing before its validity can be properly assessed.

BIOTECHNOLOGICAL DEVELOPMENT

With many crop products it is possible to translate quality into chemical terms and then begin to see how the biosynthetic pathway might be manipulated to increase the value of the product. A good example of this is found in the plant oils where the value of a particular oil for a specific application is directly related to its fatty acid constitution. Starch is another example, more comparable to cellulose, where the proportions of the branched and unbranched forms of the α-1,4-glucan polymer are important in determining the usefulness of the starch of a particular crop for a specific application. By contrast, with cellulose the quality of the crop product is related less to its chemical make-up than to the fibrillar and cellular form in which it is naturally produced. Thus in cotton, breeding has aimed to create varieties in which the fibres are uniform in length, strong and slightly coarse. These are desirable characteristics because they are important in the spinning of high quality yarns. The Sea Island cotton is famous for the fine and strong fibres, which are 50 mm long. Similarly in the manufacture of paper, long fibres are desirable because they give paper strength. However, our knowledge of cellulose biosynthesis is not yet sufficient for us to be able to identify the enzymes responsible for these quality factors.

Cellulose also differs from the other crop products in that it is accumulated in plants as part of a structure, the cell wall, which incorporates other polymers in a close association. In the case of cotton the other components of the cell wall form a very small proportion of the fibre (about 6 per cent), but the cellulose of wood fibres and plant fibres such as flax can be heavily contaminated by non-cellulosic materials, particularly lignin. Very often control of the lignin content is an important preliminary in the manufacture of a cellulose-based product, such as paper. In flax the deposition of lignin in and around the fibre bundles has long been known to have a deleterious effect on fibre quality. The effect of lignification is to delay the liberation of the fibres from the plant by retting, to cause problems with the processing the fibres, and lignification also decreases the flexibility and softness of the fibre bundles (Keijzer, 1989). Lignification occurs after the development of the fibre cells in the secondary phloem cells is otherwise complete, and therefore optimum quality depends on the stage at which the flax is pulled. The regulation of lignin biosynthesis is well understood at the biochemical level, but we are still some way from being able to formulate a programme of research which would lead to the prevention of lignification specifically of the flax fibres.

FURTHER READING

Bacic, A., Harris, P.J. & Stone, B.A. (1988). Structure and function of plant cell walls, in *The Biochemistry of Plants. Volume 14 Carbohydrates*, Ed Preiss, J., New York, Academic Press, pp. 297–371.

Delmer, D.P. & Stone, B.A. (1988). Biosynthesis of plant cell walls, in *The Biochemistry of Plants. Volume 14 Carbohydrates*, Ed Preiss, J., New York, Academic Press, pp. 373–420.

French, A.D. (1985). Physical and theoretical methods for determining the supramolecular structure of cellulose, in *Cellulose Chemistry and Its Applications*, Eds Nevell, T.P. & Zeronian, S.H., Chichester, Ellis Horwood, pp. 84–111.

Fry, S.C. (1988). *The Growing Plant Cell Wall: Chemical and Metabolic Analysis*, Harlow, Longman.

Haigler, C.H. (1985). The functions and biogenesis of native cellulose, in *Cellulose Chemistry and Its Applications*, Eds Nevell, T.P. & Zeronian, S.H., Chichester, Ellis Horwod, pp. 30–83.

Lee, J.A. (1984). Cotton as a world crop, in *Cotton*, Eds Kohel, R.J. & Lewis, C.F., Madison, American Society of Agronomy, pp. 1–25.

Nevell, T.P. & Zeronian, S.H. (1985). Cellulose chemistry fundamentals, in *Cellulose Chemistry and Its Applications*, Eds Nevell, T.P. & Zeronian, S.H., Chichester, Ellis Horwood, pp. 15–29.

Preston, R.D. (1986). Natural celluloses, in *Cellulose: Structure, Modification and Hydrolysis*, Eds Young, R.A. & Rowell, R.M., New York, John Wiley, pp. 3–27.

Ross-Murphy, S.B. (1985). Properties and uses of cellulose solutions, in *Cellulose Chemistry and Its Applications*, Eds Nevell, T.P. & Zeronian, S.H., Chichester, Ellis Horwood, pp. 202–222.

ADDITIONAL REFERENCES

Alberts, B., Bray, D., Lewis, J., Raff, M., Roberts, K. & Watson, J.D. (1983). *Molecular Biology of the Cell*. New York, Garland Publishing, Inc.

Delmer, D.P. (1987). Cellulose biosynthesis, *Annu. Rev. Plant Physiol. Plant Mol. Biol.* **38**, 259–290.

Hayashi, T. & Delmer, D.P. (1988). Xyloglucan in the cell walls of cotton fiber, *Carbohydrate Res.* **181**, 273–277.

Hayashi, T. (1990). Xyloglucans in the primary cell wall, *Annu. Rev. Plant Physiol. Plant Mol. Biol.* **40**, 139–168.

Keijzer, P. (1989). Synchronization of fibre and grain maturation of flax (*Linum usatitissimum* L.), in *Flax: Breeding and Utilisation*, Ed Marshall, G., Dordrecht, Kluwer, pp. 26–36.

Marx-Figini, M. (1969). On the biosynthesis of cellulose in higher and lower plants, *J. Polymer Sci.* Part C **28**, 57–67.

Meinert, M.C. & Delmer, D.P. (1977). Changes in biochemical composition of the cell wall of the cotton fiber during development, *Plant Physiol.* **59**, 1087–1097.

Northcote, D.H. (1984). Control of cell wall assembly during differentiation, in *Structure, Function, and Biosynthesis of Plant Cell Walls*, Ed Dugger, W.M. & Bartnicki-Garcia, S., Maryland, American Society of Plant Physiologists, pp. 222–234.

Salme'n, N.L. (1985). Mechanical properties of wood fibres and paper, in *Cellulose Chemistry and Its Applications*, Eds Nevell, T.P. & Zeronian, S.H., Chichester, Ellis Horwood, pp. 505–530.

Stewart, J. McD. (1975). Fiber initiation on the cotton ovule (*Gossypium hirsutum*), *Am. J. Bot.* **62**, 723–730.

Vian, B., Reis, D., Mosiniak, M. & Roland, J.C. (1986). The glucuronoxylans and the helicoidal shift in cellulose microfibrils in linden wood: cytochemistry *in muro* and on isolated molecules, *Protoplasma* **131**, 185–199.

Young, R.A. (1986). Structure, swelling and bonding of cellulose fibers, in *Cellulose: Structure, Modification and Hydrolysis*, Eds Young, R.A. & Rowell, R.M., New York, John Wiley, pp. 91–128.

Chapter 6

Oils

INTRODUCTION

Worldwide the production of plant oils is approximately 60 million tonnes. In terms of international commerce, plant oils are one of the most important crop products, with the value of the oil-yielding crops moving in world trade being second only to that of cereals. In the harvested crop the oil is almost invariably accompanied by a protein that provides a valuable feed for livestock. For the bulk of oil crop production, pressing or extraction separates the oil from the protein-rich cake or meal; only a relatively small proportion of the crop is used directly for food. It is this combination of two highly marketable products—the oil and the protein—which is important in the economics of production and processing of oil crops. Another important factor in the relative value of different oil-yielding crops comes from the interchangeability of different oils for the various end uses. This results in a direct competition between oils from different plant sources for a particular product. For example, oils extracted from sunflowers, soybeans and rapeseed compete with one another for the salad oil market. In addition the competitiveness of oils is affected by chemical modifications, such as hardening, which can endow an oil with attributes found naturally in another oil.

Some examples of the wide use to which plant oils are put are shown in Table 6.1. About 90 per cent of the oil extracted from crops is used for food. This is a market which has increased enormously in the past 40 years. In the populations of developed countries oil/fat provides about 35 to 45 per cent of the total energy intake, while in developing countries the comparable figure is 10 to 20 per cent. By contrast, the non-food, industrial use has remained relatively static over the same period, but recently there has been a renewed interest in plant oils for specialised non-food, industrial uses. These are discussed at the end of the chapter.

Table 6.1 Some uses of plant oils (from Hatje, 1989).

Food	Non-food
Salad oils	Pharmaceuticals
Margarine	Soaps
Shortenings	Linoleum
Cooking oils	Cosmetics
Fats for the bakery,	Lubrication
confectionery industry	Chemicals, candles
and mayonnaise manufacturers	Technical products
Oils for the fish	Plastic coatings
and canning industry	Paints and resins
Feed fats	

Oils represent a very concentrated form of energy. With 38 kJ/g, oil has a much higher density of available energy than either protein or carbohydrate (17 kJ/g). As well as supplying energy, oils provide those essential fatty acids, linoleic and linolenic acids, which mammals cannot themselves make. The oils in our food also act as carriers for the fat-soluble vitamins A,D,E and K, the absorption of which is facilitated by dietary oils. But as well as doing us good, fats and oils improve the taste of meals, making them more appetising and more satisfying.

Traditionally an important non-food use of plant oils was in the manufacture of soaps. This remains a significant use, but the properties of particular plant oils allows them to find a variety of other, more specialised roles. Thus, linseed oil reacts rapidly with oxygen to form a tough, adherent film, which makes it valuable as a drying oil in the production of paints; castor oil is important in the production of lubricants; while jojoba oil (properly not an oil, but a liquid wax) is an excellent substitute for sperm whale oil, and finds many uses in the manufacture of cosmetics.

Plant oils have been considered as fuels for diesel engines since the time of Rudolph Diesel. Compared to conventional diesel fuel, the plant oils give an overall similar performance. However, the relatively higher cost of the plant oils restricts their use to where there is a local abundance, or in an emergency. The use of plant oils as fuels need not be confined to small-scale machinery. In World War II, the boilers of the *Yamoto*, a 65 000 tonne Japanese battleship, were fuelled by soybean oil (Quick, 1989).

There are two main groups of crops that are grown for their oil (Figure 6.1): the annual (or biennial) crops, exemplified by sunflower seed and rapeseed; and the perennial trees, exemplified by the olive and the oil palm. In addition, oil is a valuable by-product of a number of crops which are not grown principally for their oil, notably corn and cotton. In some countries locally available sources are exploited for their oil. For example, in Italy oil is extracted from grape and tomato seeds. The oil content of the crops varies

Biosynthesis of Major Crop Products

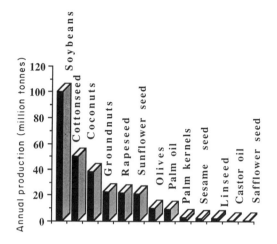

Fig. 6.1 Annual production of the major plant oils and oil-yielding crops. (Average figures for 1987–89. From FAO Production Yearbook.)

from about 35 to 50 per cent in the main oilseeds, such as sunflower, rapeseed and groundnuts, to about 18 per cent in the protein-rich soybeans.

Plants in general accumulate oils only in the seed or fruit. For the major oil crops the seed is the most important organ in which the oil is deposited, although the oil palm is an important exception, as the palm oil of commerce is extracted from the mesocarp of the fruit, while the seed yields a distinct oil, known commercially as palm kernel oil. In most seeds, including soybean, sunflower and rapeseed, oil is stored in the cotyledons of the embryo, but instead in the castor bean and coconut there is an oily endosperm.

PLANT OILS

It is convential to refer to those lipids that are liquid at ambient temperatures as oils, and to those that are solid as fats. However, even these simple definitions give us problems when the ambient temperature changes. Thus coconut oil leaves the producing countries in the tropics as an oil, and arrives in the cooler climates of Western Europe as a solid, where nevertheless it continues to be referred to as an oil (Hatje, 1989).

In chemical terms plant oils are triacylglycerols, in which fatty acids are esterified to the three hydroxy positions available on the glycerol molecule (Figure 6.2). Until recently these storage oils were described as triglycerides. This term will still be found in industry, and in the older biochemical literature, but has now been replaced by the chemically more accurate term, triacylglycerol. There is one important exception to the triacylglycerol nature of commercially exploited plant oils. That exception is the oily product of the

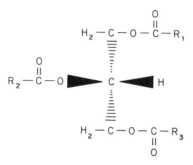

Fig. 6.2 The structure of triacylglycerols.

jojoba bean, which is actually a liquid wax made up of straight chain esters of 20:1, 22:1 and 24:1 fatty acids and fatty alcohols.

There is a wide range of fatty acids that can enter into the composition of plant oils (Table 6.2). Fatty acids with even-numbered carbon atoms predominate. Among the unsaturated fatty acids most have the double bond in the *cis*-position. In the systematic name of the unsaturated fatty acids the position of the double bond is given by the number of the first unsaturated carbon from the carboxyl end. Thus oleic acid is *cis*-9-octadecenoic acid. The symbol indicates the number of carbons in the chain and the number of unsaturated bonds. It can also include in parenthesis the number of carbon atoms from the methyl end to the first double bond. Thus oleic acid has the symbol 18:1 (n-9).

The polyunsaturated fatty acids, particularly linoleic and linolenic acids, are important constituents of plant oils. In these molecules the double bonds are separated by a single methylene ($-CH_2-$) group. There are two isomers of linolenic acid (Figure 6.3). The common form is the α-isomer, which is widely distributed; the more unusual form, γ-linolenic is found in a restricted group of seeds, which includes evening primrose, borage and currents.

It is the nature of the fatty acids present that gives a particular oil its characteristic properties, and thus determines its nutritional value and indust-

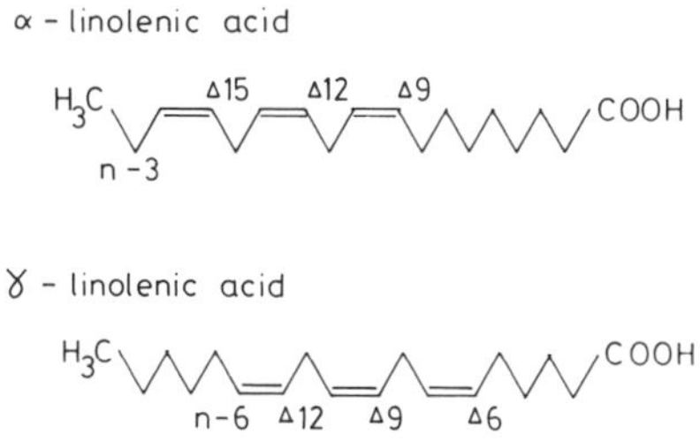

α - linolenic acid

γ - linolenic acid

Fig. 6.3 The structures of the two isomers of linolenic acid.

Table 6.2 Fatty acids present in plant oils.

Systematic name	Trivial name	Symbol
Saturated		
Decanoic	Capric	10:0
Dodecanoic	Lauric	12:0
Tetradecanoic	Myristic	14:0
Hexadecanoic	Palmitic	16:0
Octadecanoic	Stearic	18:0
Eicosanoic	Arachidic	20:0
Mono-unsaturated		
Cis-9-dodecenoic	Lauroleic	12:1 (n-3)
Cis-9-tetradecenoic	Myristoleic	14:1 (n-5)
Cis-9-hexadecenoic	Palmitoleic	16:1 (n-7)
Cis-9-octadecenoic	Oleic	18:1 (n-9)
Cis-9-docosenoic	Erucic	22:1 (n-9)
Poly-unsaturated		
Cis-cis-9,12-octadecadienoic	Linoleic	18:2 (n-6)
All-cis-9,12,15-octadecatrienoic	α-Linolenic	18:3 (n-3)
All-cis-6,9,12-octadecatrienoic	γ-Linolenic	18:3 (n-6)

rial application. The main commercial oils contain predominantly palmitic, oleic and linoleic acids (Figure 6.4). However, some oils find an industrial or nutritional application because of the presence of certain other fatty acids (Table 6.3).

Within each triacylglycerol molecule the three fatty acids may be similar, like the trioleoylglycerol of olive oil, or different, as in the 1-palmitoyl, 2-oleoyl, 3-stearoylglycerol of cocoa butter. In general, plant oils are a complex mixture of a large number of different triacylglycerols, but there is a prefer-

Table 6.3 Major source and applications of oils rich in particular fatty acids.

Fatty acid	Major source	Applications
Short chain		
Lauric 12:0	Coconut	Soaps, detergents
	Palm kernel	
Medium chain, polyunsaturated		
Linolenic 18:3	Linseed	Drying oil
Hydroxy		
Ricinoleic	Castor bean	Lubricant, paints
Very long chain		
Erucic (22:1)	Rapeseed	Lubricant

ence for an unsaturated acid to occupy the *sn*-2 position, and where saturated acids are present, they show a preference for the *sn*-1 and *sn*-3 positions.

On extraction, plants oils contain a variety of minor constituents that can affect the stability and quality of the oil (Table 6.4).

The free fatty acids, mono- and diacyl glycerols originate either as intermediates in triacylglycerol biosynthesis, or from the partial hydrolytic break-

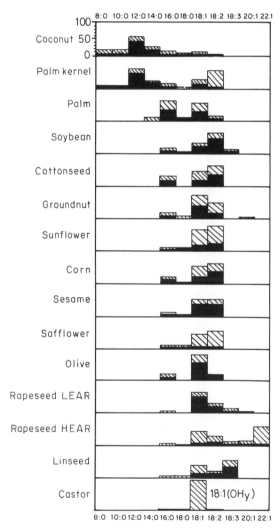

Fig. 6.4 Percentage fatty acid composition of plant oils. The maximum (cross-hatched) and minimum (filled) contents for different varieties are shown. For rapeseed, high (HEAR) and low (LEAR) erucic acid types are shown separately. (Data from Gunstone *et al.*, 1986.)

Table 6.4 Minor constituents of extracted plant oils.

Free fatty acids, mono- and diacylglycerols
Phospholipids and glycolipids
Sterols
Waxes
Carotenoids and chlorophylls
Tocopherols

---

down of the oil by lipases. Consequently the presence of these compounds in seed oils indicates that the seeds were immature at harvest, or that they have been mechanically damaged, or that they have started to germinate. The phospholipids and glycolipids originate in large part from the membranes of chloroplasts and other cell organelles. They appear in greater amounts in the oil when the oil is extracted at higher temperatures and with harsher extraction procedures. Phospholipids and glycolipids are not necessarily undesirable components of the oil, however, as collectively they are recovered from the oil as a 'lecithin' fraction to be used in the food industry as emulsifiers. Chemically, of course 'lecithin' is the name for only one particular phospholipid, phosphatidyl choline. The waxes found in plant oils originate from the seed coat surface. They are generally separated from the oil proper as a solid fraction when the crude oil is cooled. A variety of sterols may be present in plant oils. While some are widespread, others are characteristic of the species from which the oil was extracted, so that they can be used to identify the origin of the oils that have been incorporated in a blend. The most notable example is brassicasterol, the presence of which can be used to identify oils from rapeseed and other plants belonging to the Cruciferae. The chlorophyll and carotenoid pigments contribute to the colour of the crude oil, which is generally amber, but can be green, as in olive oil, or red, as in palm oil. The tocopherols in oil play an important part in maintaining oil quality during storage, because they slow the rate at which unsaturated fatty acids in the oil are oxidised on exposure to air.

Within the cells of the storage organ the oil is deposited as droplets (Figure 6.5). In the past these have been given various names in the literature: spherosomes, oil bodies, reserve oil droplets and lipid-protein particles. Currently, the term which is perhaps most widely used is oil bodies. Electron-microscopy reveals that oil bodies first appear as naked droplets in the cytosol (Figure 6.5A). Later in the development of the seed they acquire an osmiophilic membrane. In rapeseed, where they have been most intensively studied (Murphy *et al.*, 1989), the immature oil droplets contain little or no protein, but do contain some phospholipid. It is only towards the end of embryogenesis, when the bulk of the oil and the storage protein has been synthesised that they acquire the osmiophilic layer (Figure 6.5C,D). At this later stage the protein content of the oil droplets rises to constitute about 20 per cent by weight of the total seed protein (Figure 6.6).

Oils   95

Most of the oil droplet protein consists of a single molecular species: a hydrophobic polypeptide of 19 kDa termed oleosin (or olein). The oleosins are of widespread occurrence in the Cruciferae, and comparable hydrophobic proteins occur in other non-cruciferous species. Their function is not yet clear, but the timing of their appearance relatively late in seed development argues against a role for them in the synthesis of oil. What is more likely is that they act as emulsifiers, helping to stabilise the oil bodies during the subsequent desiccation and rehydration stages of seed development.

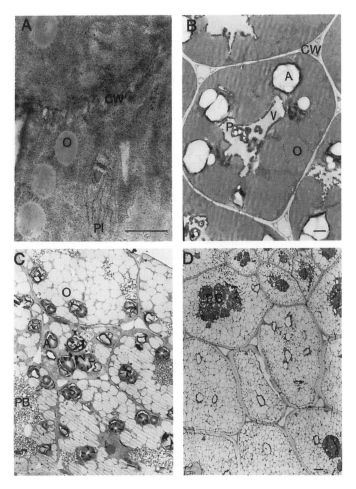

Fig. 6.5   Oil bodies in the developing embryos of rapeseed. Samples were taken for electron-microscopy at (A) 2 weeks, (B) 4 weeks, and (C) 6 weeks post anthesis, and (D) after the seed had dried. The bars represent 1μm. A, amyloplast; CW, cell wall; O, oil body; PB, protein body; Pl, plastid; V, vacuole. (Courtesy of Dr D.J. Murphy. Reproduced by permission of The Biochemical Society.)

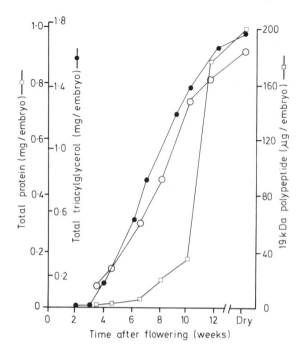

Fig. 6.6   Appearance of the 19 kDa oil-body polypeptide during the development of rapeseed. (Redrawn from Murphy *et al.*, 1989. Reproduced by kind permission of the authors and The Biochemical Society.)

BIOCHEMISTRY OF PLANT OIL SYNTHESIS

The primary substrate for oil synthesis is the sucrose transported from the photosynthetic tissues to the oil synthesising organs. This sucrose supplies both the glycerol and the fatty acids of the oil molecule. The sucrose is initially broken down by invertase or sucrose synthase, and the resulting hexose units are metabolised via glycolysis to generate the basic building blocks of fatty acid synthesis, the acetyl-CoA units.

The subsequent synthesis of the fatty acids from acetyl-CoA can then be considered to occur in three stages (Figure 6.7). First, the acetyl units undergo a series of condensation reactions to form palmitic acid (16:0) and the medium chain (10:0–14:0) saturated fatty acids. Second, the palmitic acid is modified by elongation, unsaturation and other reactions to form the full range of fatty acids. Third, the fatty acids are esterified to the glycerol backbone to form the triacylglycerol product. As we shall see later, the second and third stages overlap to some extent, since the fatty acids can undergo modifying reactions, such as desaturation, after they have been added to glycerol or phospholipid.

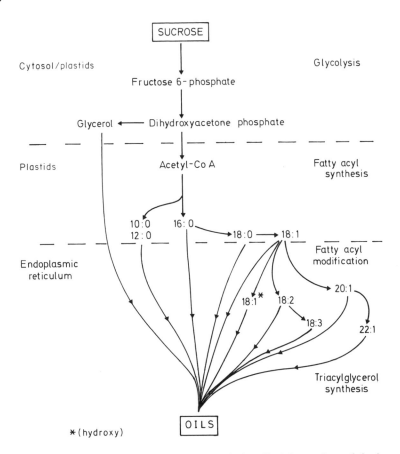

Fig. 6.7 Outline of triacylglycerol synthesis in oil-rich seeds and fruits.

Synthesis up to the stage of oleic acid is confined to plastids—chloroplasts in the leaves, and proplastids in developing oilseeds and mesocarps of oil-yielding fruits. Subsequent modification of the oleic acid occurs outside the plastid, generally on the endoplasmic reticulum, which is also the intracellular location of triacylglycerol formation.

Synthesis of palmitate from acetyl-CoA

The acetyl-CoA can be derived either from acetate or from pyruvate. The formation of acetyl-CoA from acetate is catalysed by acetyl-CoA synthetase, which is an enzyme widely distributed in plants:

$$\text{acetate} + \text{ATP} + \text{CoA} \rightarrow \text{acetyl-CoA} + \text{AMP} + \text{PP}_i$$

This reaction is useful experimentally because it allows radioactively labelled acetate, which is readily available, to be used as a labelled substrate for fatty acid synthesis. However, it is not clear how the acetate would be produced *in vivo*.

The pyruvate is generated from sucrose by glycolysis. There is now evidence that plastids contain all the glycolytic enzymes, as well as the enzymes of the oxidative pentose phosphate pathway (Stumpf, 1989). Thus not only can plastids generate all the cofactors required for fatty acid synthesis: ATP, NADPH and NADH, but they also possess the enzymes required for the conversion of hexose units to fatty acids. Thus pyruvate can be produced either within the fatty-acid synthesising plastids or from cytosolic glycolysis. Acetyl-CoA is generated from pyruvate by an oxidative decarboxylation catalysed by the pyruvate dehydrogenase complex.

$$\text{pyruvate} + \text{CoA} + \text{NAD} \rightarrow \text{acetyl-CoA} + CO_2 + \text{NADH}$$

The elongation of the fatty acids occurs by the sequential addition of two-carbon acetyl units. These are derived from acetyl-CoA but they are donated to the growing fatty acid chain as part of the malonyl-CoA molecule (Figure 6.8). Thus the synthesis of malonyl-CoA by the carboxylation of acetyl-CoA catalysed by acetyl-CoA carboxylase is referred to as the first committed step in the synthesis of fatty acids, although it also provides two-carbon units for the synthesis of flavonoids and cuticular waxes. The levels of this enzyme in developing rape seeds increase just before the onset of the grand phase of lipid accumulation, and it is possible that it has a critical role in determining the amount of oil accumulated by commercially important oilseed crops (Hellyer *et al.*, 1986).

The acetyl-CoA carboxylase of plants, including rapeseed, contains biotin, and as in animals and bacteria, catalyses a 2-step reaction:

(1)   biotin carboxylation

$$\text{enzyme-biotin} + \text{ATP} + HCO_3^- \rightarrow \text{enzyme-biotin-}CO_2 + \text{ADP} + P_i$$

(2)   transcarboxylation

$$\text{enzyme-biotin-}CO_2 + \text{acetyl-CoA} \rightarrow \text{malonyl-CoA} + \text{enzyme-biotin}$$

The acetyl-CoA carboxylase of rapeseed, with a relative molecular mass of 220 kDa seems to resemble more closely the enzyme of animals and fungi, which consists of a single relatively large polypeptide, than that of bacteria, where the biotin carboxyl carrier protein is a separate entity from the other two polypeptides, which constitute the carboxylase and transcarboxylase enzymes (Hellyer *et al.*, 1986).

Synthesis of the fatty acids of plant oils occurs, as in animals and bacteria, while the intermediates are covalently linked to a small (10 kDa), heat-stable protein called the acyl carrier protein (ACP). Attachment to the ACP occurs

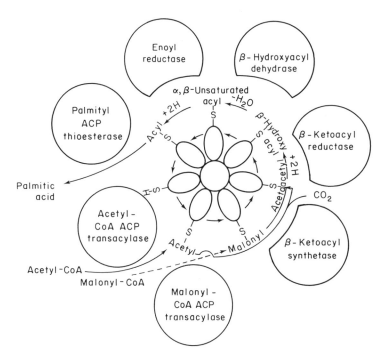

Fig. 6.8 Schematic representation of fatty acid synthetase from an oil-accumulating seed or fruit. Elongation of the acyl group takes place while it is attached to the acyl carrier protein.

via 4′-phosphopantetheine, the free end of which is terminated by a sulphydryl-group, the other end being attached to a serine reside of the ACP. Acyl groups are also attached to CoA via 4′-phosphopantetheine, and the function of ACP during synthesis resembles that of CoA during fatty acid oxidation. In animals and yeast the ACP is tightly associated with the enzymes responsible for fatty acid biosynthesis to form a multienzyme complex, individual enzyme components of which are largely inactive when isolated. This is designated a Type I fatty acid synthase system. By contrast the plant fatty acid synthase system resembles the bacterial type, where the biosynthetic enzymes catalysing the partial reactions are normally in a dissociated form. In this Type II synthase each enzyme remains active when it is separated from the others. It may be significant that in animals and yeast fatty acid synthesis occurs in the cytosol, whereas in plants it is confined to the plastids, which resemble eubacteria in being essentially prokaryotic in organisation. In developing rapeseed the molecular cloning of cDNA encoding the ACP has revealed the presence of a multigene family encoding five different mature ACP polypeptides (Safford *et al.*, 1988). These are not expressed in the leaf (where other ACP genes are expressed), and thus are likely to be specific to the biosynthesis of the seed oils. Despite the plastid localisation, ACP is encoded in the nucleus, the precursor molecule possessing an N-terminal extension, which probably functions as a transit peptide to direct the newly synthesised ACP into the plastid.

The initial step in the elongation of the fatty acids is the transfer of acetyl and malonyl groups to the ACP. These transfers are catalysed by the respective acetyl and malonyl transacylases. The reaction between malonyl-ACP and acetyl-ACP now takes place catalysed by β-ketoacyl-ACP synthase. In this condensation reaction the carbon dioxide that was fixed in the synthesis of malonyl-CoA is released. The ATP consumed in the synthesis of malonyl-CoA has in effect driven the condensation reaction, although it has done so indirectly by providing the free energy for the synthesis of the malonyl group. The decarboxylation of the malonyl group has in turn favoured energetically the condensation reaction, so that formation of the acetoacetyl group takes place.

The subsequent steps in each turn of the cycle of synthesis have the net effect of reducing the keto group of the added acetyl residue to a methylene group (Figure 6.9). First there is a reduction to form the β-hydroxy fatty acyl-ACP, then a dehydration to form an α, β-unsaturated fatty acyl-ACP, and a

Table 6.5 Enzymes that have been identified as being involved in the synthesis of oil from acetyl-CoA.

1. Acetyl-CoA carboxylase
    Acetyl-CoA:carbon dioxide ligase (ADP forming)   EC 6.4.1.2

    acetyl-CoA + CO_2 + ATP $\rightarrow$ malonyl-CoA + ADP + P_i

2. Acetyl-CoA:ACP transacylase
    Acetyl-CoA:ACP S-acetyltransferase   EC 2.3.1.38

    acetyl-CoA + ACP $\longleftrightarrow$ acetyl-ACP + CoA

3. Malonyl CoA:ACP transacylase
    Malonyl-CoA:ACP S-malonyltransferase   EC 2.3.1.39

    Malonyl-CoA + ACP $\longleftrightarrow$ malonyl-ACP + CoA

4. β-ketoacyl-ACP synthase I
    Acyl-ACP:malonyl-ACP C-acyltransferase (decarboxylating)   EC 2.3.1.41

    Acyl-ACP(C2–C14) + malonyl-ACP $\rightarrow$ β-ketoacyl-ACP (C4–C16)
    $+$ CO_2 + ACP

5. β-ketoacyl-ACP synthase II
    Acyl-ACP:malonyl-ACP C-acyltransferase (decarboxylating)   EC 2.3.1.41

    Palmitoyl-ACP + malonyl-ACP $\rightarrow$ β-ketostearoyl-ACP + CO_2 + ACP

Table 6.5 (*continued*)

6. β-ketoacyl-ACP reductase
   3-hydroxyacyl-ACP:NADP$^+$ oxidoreductase   EC 1.1.1.100

   β-ketoacyl-ACP + NADPH $\rightarrow$ D-β-hydroxyacyl-ACP + NADP

7. D-β-hydroxyacyl-ACP dehydratase
   3-hydroxypalmitoyl-ACP hydrolyase   EC 4.2.1.61

   D-β-hydroxyacyl-ACP $\longleftrightarrow$ *trans*-2-enoyl-ACP + H$_2$0

8. Enoyl-ACP reductase
   Acyl-ACP:NAD$^+$ oxidoreductase   EC 1.3.1.9

   2-enoyl-ACP + NADH $\longleftrightarrow$ acyl-ACP + NAD

9. Stearoyl-ACP desaturase
   Acyl-ACP, hydrogen donor:oxygen oxidoreductase   EC 1.14.99.6

   stearoyl-ACP + AH$_2$ + O$_2$ $\rightarrow$ oleoyl-ACP + A + 2H$_2$O

10. Glycerol 3-phosphate acyltransferase
    Acyl-CoA:*sn*-glycerol 3-phosphate *O*-acyltransferase   EC 2.3.1.15

    glycerol 3-phosphate + acyl-CoA $\rightarrow$ 1-acylglycerol 3-phosphate + CoA

11. 1-acylglycerol-3-phosphate acyltransferase
    Acyl-CoA:1-acyl-*sn*-glycerol-3-phosphate *O*-acyltransferase   EC 2.3.1.51

    1-acylglycerol 3-phosphate + acyl-CoA $\rightarrow$ 1,2-diacylglycerol 3-phosphate + CoA

12. Phosphatidate phosphatase
    3-*sn*-phosphatidate phosphohydrolase   EC 3.1.3.4

    phosphatidate + H$_2$O $\rightarrow$ 1.2-diacylglycerol + P$_i$

13. Diacylglycerol acyltransferase
    Acyl-CoA:1,2-diacylglycerol *O*-acyltransferase   EC 2.3.1.20

    1,2-diacylglycerol + acyl-CoA $\rightarrow$ triacylglycerol + CoA

14. 1-Acylglycerophosphocholine acyltransferase
    acyl-CoA:1-acyl-*sn*-glycero-3-phosphocholine *O*-acyltransferase   EC 2.3.1.23

    1-acylglycero-3-phosphocholine + acyl-CoA $\leftrightarrow$ 1,2-diacylglycero-3-phospho-choline + CoA

15. Cholinephosphotransferase
    CDPcholine:1,2-diacylglycerol cholinephosphotransferase   EC 2.7.8.2

    CDPcholine + 1,2-diacylglycerol $\rightarrow$ CMP + a phosphatidylcholine

Biosynthesis of Major Crop Products

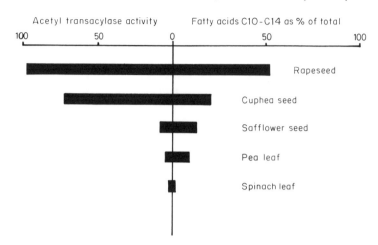

Fig. 6.9 Relationship between the proportion of medium-chain fatty acids produced by plant extracts and the acetyl transacylase activity of the extracts. (Data from Shimakata & Stumpf, 1983.)

further reduction to a saturated fatty acyl-ACP. The condensation of this product with malonyl-ACP initiates a further turn of the synthetic cycle. With seven turns of the cycle palmitoyl-ACP is formed.

The further elongation to stearoyl-ACP depends on the activity of a specific Type II β-ketoacyl-ACP synthase, which is inactive with shorter chain fatty acyl-ACPs. The normal reductant for the β-ketoacyl reductase and enoyl reductase is NADPH. However, in developing safflower seeds one of the two types of enoyl reductases reduces only the 4-carbon crotonyl-ACP, and uses NADH as reductant.

In oil seeds and fruits the final products of the fatty acid synthase are generally the 16:0 and 18:0 fatty acids, except, as in the endosperms of palm kernel and coconut, where the medium-chain fatty acids (C10–C14) are major products. Despite the commercial importance of these oils, the mechanism of chain termination remains unknown. It may be that specific β-ketoacyl thioesterases limit further chain elongation by catalysing the release of a specific chain from the ACP, or specific β-ketoacyl thioester synthases catalyse the transfer of specific chains to CoA for subsequent transfer to triacylglycerol. However, an indication of how the relative activities of the enzymes of the fatty acid synthase system could affect the fatty acid mix produced by the fatty acid synthase system has come from experiments in which the levels of each of the constituent enzymes were varied in a reconstituted system (Shimakata & Stumpf, 1983). This kind of experiment is possible with the plant fatty acid synthase system, because the component enzymes are not tightly associated, as they are in the animal and yeast systems.

When ^{14}C-labelled malonyl-CoA was supplied to crude extracts prepared from different plants, the saturated fatty acids synthesised ranged from 10:0

to 18:0. Conditions were unsuitable for their further modification to longer chain and unsaturated fatty acids. The major products were stearic and palmitic acids, as expected, but medium-chain (C10–C14) fatty acids were also produced (Figure 6.9). When the seven enzymes of the fatty acid synthase system were assayed in the crude extracts, the activities of acetyl transacylase and β-ketoacyl-ACP synthases I and II were significantly lower than those of the other enzymes. Moreover there appeared to be a relationship between the activity of the acetyl transacylase and the proportion of medium-chain fatty acids formed (Figure 6.9), so that rapeseed extract with an active acetyl transacylase produced the highest proportion of medium-chain fatty acids, and the spinach extract with a relatively inactive acetyl transacylase the lowest proportion. Addition of increasing amounts of a purified acetyl transacylase to the spinach extract increased the proportion of medium-chain fatty acids produced. Moreover, when the levels of acetyl transacylase were increased in a reconstituted system of purified enzymes and ACP, there was a corresponding increase in the proportion of medium-chain length fatty acids in the product formed from acetyl-CoA and ^{14}C-labelled malonyl CoA. By contrast increasing the levels of the other enzymes in the reconstituted system simply increased the rate of formation of the final product, stearic acid. This work of Shimikata & Stumpf (1983) provides evidence that the acetyl transacylase is not only a rate-limiting enzyme, but also that its activity relative to that of the other enzymes can control the nature of the fatty acids produced. Presumably a relatively high activity of the acetyl transacylase generates a large population of acetyl-ACPs, which successfully compete for malonyl-CoA with the β-ketoacyl synthase I, which initiates each turn of the cycle of chain elongation. These experiments show us how fatty acid chain length and enzyme activity can be correlated *in vitro*, but the significance of a mechanism such as this for the regulation of chain length *in vivo* remains unknown.

Formation of unsaturated fatty acids

Stearic acid forms the substrate for the synthesis of oleic acid. The desaturation is catalysed by the enzyme stearoyl-ACP desaturase. This enzyme has been purified and shown to require both molecular oxygen and a source of reducing electrons, preferably reduced ferredoxin (Figure 6.10). Presumably in the non-photosynthetic plastids of developing oilseeds the ferredoxin is reduced by NADPH, the reaction being catalysed by ferredoxin-NADP oxidoreductase. Oleic acid is released on hydrolysis of the ACP complex by a specific acyl-ACP thioesterase. Unfortunately the desaturase is unstable *in vitro* and the reaction remains poorly understood.

The elongation of oleic acid probably occurs via the successive addition of two-carbon units from malonyl-CoA to oleoyl-CoA. Microsomal fractions capable of supporting these reactions have been prepared from a number of species belonging to the Cruciferae, where erucic acid (22:1) is an important component of the seed oil.

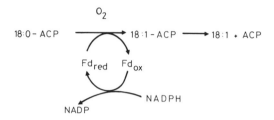

Fig. 6.10 Aerobic synthesis of oleic acid from stearoyl-ACP using ferre-doxin (Fd).

In castor beans the predominant fatty acid is ricinoleic acid (18:1 hydroxy). There is a microsomal hydroxylase capable of introducing the hydroxyl-group at the carbon-12 of oleic acid. The substrate of the reaction is not known, but may well be an oleoyl-phosphatidylcholine.

In the synthesis of jojoba oil (Figure 6.11), the oleoyl-CoA undergoes a series of 2-carbon additions from malonyl-CoA followed by reductive steps to form the corresponding long-chain alcohols. The long-chain acids and alcohols so formed then undergo condensation reactions to form the characteristic wax esters of jojoba oil. Again details of the reactions involved are not known.

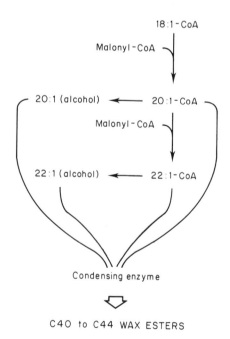

Fig. 6.11 Scheme for the synthesis of jojoba esters.

Formation of triacylglycerols

Oils are one of the few crop products in which biosynthesis can actually be observed to take place 'in the test tube'. Stobart *et al.* (1986) have shown that microsome preparations derived from developing cotyledons of safflower are capable of rates of triacylglycerol formation from acyl-CoA and glycerol phosphate which are comparable to the rates of oil synthesis *in vivo*. The gradual development of oil droplets can be observed with the electron microscope after 15 minutes incubation. After 1 hour the reaction mixture acquires a milky appearance, and continued incubation leads to the solution becoming opaque with the accumulation of oil droplets. A demonstration such as this is possible because the enzymes responsible for catalysing the synthesis of the triacylglycerol from the glycerol backbone and the acyl-CoA units are intrinsic proteins of the endoplasmic reticulum and therefore carried over undamaged in the preparation of the microsomes.

In oil storage tissues palmitate, stearate, and oleate are transferred from the plastid for further metabolism in the cytoplasm. The mode of transport through the plastid membrane has not yet been described, although it is clear that subsequent metabolism in the cytoplasm requires the fatty acid to be esterified to CoA.

The glycerol backbone of the triacylglycerol is derived from the reduction of dihydroxyacetone phosphate by a reductase located in the cytosol. The resulting glycerol 3-phosphate now enters the glycerol phosphate (or Kennedy) pathway, which is shown in Figure 6.12.

In this pathway acyl groups are added successively to the three positions of the glycerol 3-phosphate. In most of the commercially important oilseeds, linoleate makes an important contribution to the fatty acid composition, and in linseed oil, linolenate is an important constituent (Figure 6.4). These unsaturated fatty acids are derived by desaturation reactions from oleic acid while they are esterified to phosphatidyl choline. Thus there are additional biosynthetic steps that allow acyl exchange to occur between acyl groups esterified to phosphatidyl choline and those destined for inclusion in the triacylglycerol product. This exchange allows enrichment of the diacylglycerol with polyunsaturated acyl components that are formed by desaturation of oleoyl groups on phosphatidyl choline. The exchange also provides a means whereby oleoyl groups can be transferred to phosphatidyl choline for desaturation. Figure 6.12 illustrates the reactions involved in the synthesis of triacylglycerols in a linoleate-rich oil, such as that of safflower seed. A corresponding pathway is likely to be followed in the synthesis of the triacylglycerols of a linolenate-rich oil such as linseed.

Acylation at the position *sn*-1 of the glycerol 3-phosphate occurs with a preference for the saturated fatty acids, while the subsequent acylation at the *sn*-2 position shows a preference for linoleate, effectively excluding entry of saturated fatty acids at this position. These positional preferences are attributable to the specificities of the two enzymes involved: the glycerophosphate acyltransferase and the 1-acylglycerol 3-phosphate acyltransferase.

Biosynthesis of Major Crop Products

Linoleate-CoA becomes available in the cytoplasmic acyl-CoA pool following exchange between cytoplasmic oleoyl-CoA and linoleate formed by desaturation on endoplasmic reticulum-bound phosphatidylcholine.

Loss of the phosphate group from the phosphatidic acid is catalysed by a specific phosphatase to yield the diacylglycerol. By a reversible reaction catalysed by a CDPcholine:diacylglycerol cholinephosphotransferase, the

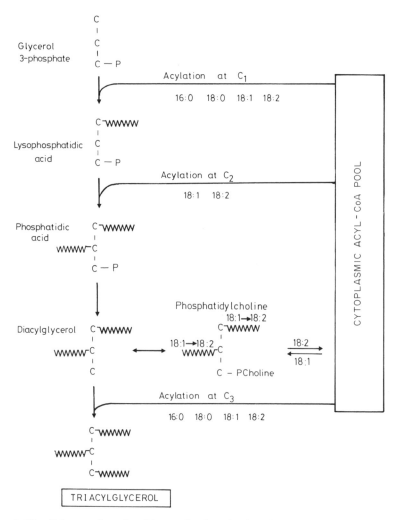

Fig. 6.12 Scheme for the biosynthesis of triacylglycerols in linoleate-rich oilseeds. Enrichment of the cytoplasmic pool of linoleate-CoA occurs after desaturation of oleate to linoleate on positions 1 and 2 of phosphatidylcholine. The linoleate-CoA then becomes available for acylation at positions 1,2 and 3 of the triacylglycerol (for details see Stymne & Stobart, 1987).

diacylglycerol can enter and leave the pool of phosphatidylcholine, thus allowing a further opportunity for the oleoyl groups to become desaturated. Exchange between the acyl groups of the acylphosphatidylcholine and those of the cytoplasmic pool is catalysed by the action of the enzyme acyl-CoA:lysophosphatidylcholine acyltransferase. This enzyme catalyses a reversible transfer of acyl groups between an acylphosphatidylcholine and an acyl-CoA. Exchange is brought about when both forward and reverse reactions are occurring with different acyl groups. For example, the incorporation of an oleoyl group in exchange for a linoleate group would occur as follows:

First, a linoleoyl group is taken from the *sn-2* position of the acylphosphatidylcholine:

1,2-linoleoyl-phosphatidylcholine + CoA → 1-linoleoyl-lysophosphatidy-
lcholine + linoleoyl-CoA

Then a re-acylation of the lysophosphatidylcholine produced at the *sn-2* position inserts an oleoyl group:

1-linoleoyl-lysophosphatidylcholine + oleoyl-CoA → 1-linoleoyl, 2-oleoyl-
phosphatidylcholine + CoA

An interesting enzyme specificity of these exchanges has been revealed by the *in vitro* studies. In microsomes from both safflower and linseed there was a marked preference for the transfer of oleate and linoleate to phosphatidylcholine. Acylation of linolenate occurred at about half the rate, but the saturated fatty acids, palmitate and stearate were essentially excluded from acylation when the unsaturated C18 fatty acids were excluded (Stymne & Stobart, 1987).

The oleate and linoleate desaturates are known to require O_2, and NADH or NADPH, and are probably located in the endoplasmic reticulum. However, little more is known of their properties. One reason for this is that no cell-free systems capable of purification have been obtained. Thus the enzymes are poorly characterised and the biosynthetic mechanism is not understood. This is unfortunate as the effectiveness of these desaturases in converting oleic to linoleic and linolenic acids is likely to be an important factor, if not the controlling factor, in regulating the relative proportion of the mono- and polyunsaturated fatty acids in the plant oil.

The biochemistry of those plants, such as evening primrose, borage and currants, that incorporate γ-linolenate into their storage oils, has not been examined in detail. However, the evidence that is available suggests that the desaturase enzyme which introduces the third double bond into linoleic acid acts upon a 2-linoleoyl-phosphatidylcholine substrate (Stymne & Stobart, 1987). It is also possible that the final step in biosynthesis, transfer of a third acyl group from the cytoplasmic pool of acyl-CoA, yields the triacylglycerol product. This acylation shows a similar specificity towards the acyl-CoA as the acylation at the *sn-1* position, so that this acylation provides a further opportunity for the entry of saturated fatty acids to the triacylglycerol (Figure 6.12). Stymne & Stobart (1987) point out that the potential importance of the reaction responsible for the acylation of the diacylglycerol should not be underrated. Not only is it responsible for a third of the fatty acids present in the plant oil, but this reaction is the only one which is unique to the biosynthesis of triacylglycerols, since the other acylation steps will also occur in the synthesis of structural lipids, such as membrane phospholipids. It is therefore an important enzyme in regulating the acyl quality of the triacylglycerol in plant storage oils.

FUTURE DEVELOPMENTS

The oilseeds that are currently grown on a large scale are the products of past plant breeding. This effort has resulted in varieties which combine high yields of oil or protein with disease resistance and other advantageous agronomic characteristics. Improvements in the composition of the oil have also been accomplished. These have sometimes taken place alongside useful changes in plant architecture. For example the sunflower has benefited from increases in the linoleic acid content, and also by the elimination of branches on the flowering axis so that flowers are consolidated into a single head. Complementing the activities of the plant breeder, the chemist has learnt to manipulate the quality of the extracted oil through blending, rearrangement of the fatty acids on the glycerol, and by hydrogenation. However, the future pattern of production is likely to be shaped principally by the ability of recombinant DNA techniques to create new genotypes determined by the demands of the consumer. As the following examples illustrate, development in this direction is underpinned by a knowledge of the key biochemical steps that regulate plant oil biosynthesis, in particular those metabolic events that determine the types of fatty acids in the oil produced (Table 6.6).

A major achievement of the plant breeders has been the creation of varieties of rapeseed which produce an oil containing virtually no erucic acid. Rapeseed oil, like oil from other members of the Cruciferae, normally contains 20–40 per cent erucic acid. This acid is claimed to be damaging to the membranes of animal cells, and therefore it once severely limited the market for rapeseed oil as an edible oil. The new low erucic varieties, especially when coupled to a low glucosinolate content in the protein meal, quickly replaced the traditional varieties, and have now made rapeseed the principal oilseed crop of temperate zones. The decreased content of erucic acid is associated with a correspondingly increased content of the nutritionally desirable oleic acid, and there is no loss of yield. Presumably the biochemical explanation of this substitution is that in the new varieties there is a deficiency in the elongase enzymes responsible for the conversion of oleoyl-CoA to eruceoyl-CoA.

Undesirable though the erucic acid may be in edible oils, it nevertheless finds a specialised use in industry as a feedstock for the manufacture of

Table 6.6 Desired alterations in plant oils in relation to the enzymes of biosynthesis.

Crop	Innovation	Key enzyme
Oilseed rape (HEAR)	Higher erucic content (>66%)	Specific acyltransferase (from the garden nasturtium?)
Oilseed rape (LEAR)	Lower linolenic	Decreased linoleic desaturase
Temperate oilseed crops	Accumulation of oil rich in lauric	Specific medium-chain acyl-ACP thioesterase (from other spp.?)

HEAR, high erucic acid rapeseed; LEAR, low erucic acid rapeseed.

lubricants, nylon and plasticisers. This market is supplied by varieties of rapeseed which contain high erucic acid levels. In the USA, crambe, a member of the Cruciferae with an oil containing 55–60 per cent erucic acid, has been introduced as a novel oilseed crop to supply the domestic requirement for erucic acid. In developing the high erucic acid varieties of oilseed rape it was observed that there is an upper limit of 66 per cent to the content of erucic acid, which is imposed by the restriction of erucic acid to the *sn*-1 and *sn*-3 postitions of the glycerol backbone. However, there does not appear to be an insuperable barrier to entry of erucic acid at the *sn*-2 position, since the seeds of the garden nasturtium with an erucic acid content of 80 per cent, contain 1,2,3-trieruceoyl glycerol. Presumably the specificities of the acyltransferase enzymes in nasturtium and rapeseed differ with respect to the position on the glycerol molecule to which they are capable of donating the eruceoyl group from eruceoyl-CoA. Recognition of the availability of the appropriate enzyme activity in nasturtium has raised the prospect of gene transfer of this activity directly to rapeseed in order to raise the potential erucic acid content.

Oils rich in lauric acid (12:0) provide the basis for the manufacture of soaps and detergents, and consequently have the largest market for all the non-edible uses of plant oils. The laureate-rich oils are extracted mainly from two sources: coconuts and oil palm kernels. Following a search for alternative sources in the USA, the wild species cuphea has recently been introduced as a novel oilseed crop. Unusually for a temperate annual, its oil is rich in medium-chain fatty acids and contains about 40 per cent lauric acid. However, as a recent introduction from the wild, this species suffers from problems of seed shattering, indeterminacy, and seed dormancy. Therefore as an alternative to the domestication of cuphea, another approach to the development of alternative sources of a laureate-rich oil would be to alter genetically the composition of the oil produced by a relatively productive, well-established crop plant such as oilseed rape so that the oil produced contained lauric acid instead of the oleic, linoleic and linolenic acids (Knauf, 1987). Oilseed rape also has the advantage of being relatively easily manipulated genetically and it has a short life cycle, at least in comparison with the coconut and oil palm. Laureate is an intermediate in the biosynthesis of the longer chain (C18) fatty acids made by the fatty acid synthase. Therefore the key enzymatic step for its incorporation into the triacylglycerol product would be the release of lauric acid from the ACP of the fatty acid synthase system. The enzyme responsible for this release in coconut and oil palm has not been identified, but in rat mammary tissue further elongation of the C12 and C14 acyl chains is prevented by the activity of an acyl-ACP thioesterase II. When this enzyme is purified and added to plant extracts it remains effective in removing ACP from the elongating acyl chains. The problems involved in incorporating by gene transfer this enzyme into a plant such as oilseed rape should not be underestimated however. To function in its new host the active form of the acyl-ACP thioesterase would have to be transported into the plastid (the rat enzyme is located in the cytoplasm); expression would have to be restricted to the developing embryo tissue by an appropriate tissue-specific promoter; and the level of gene expression would have to be strictly controlled so that sufficient lauric acid was produced to make a viable product without restricting the synthesis of the longer-chain fatty acids required for membrane lipids and thus for proper seed development.

Hydroxylated oils are also used on a large scale by industry for lubrication and in manufacturing plastics and cosmetics. The predominant hydroxylated fatty acid is ricinoleic acid (C18:1, hydroxy) obtained from the castor bean. Members of the Umbelliferae, such as carrot, produce an oil containing another hydroxy acid, petroselenic acid, formation of which depends on a $\triangle 6$ desaturation of stearic acid, rather than the $\triangle 9$ desaturation which leads to oleic and ricinoleic acids. Members of the Umbelliferae cannot be cultivated as high yielding oil-seed crops. Therefore to meet the industrial demands for petroselenic acid an established oilseed crop such as rape would need to be transformed with genes for the $\triangle 6$ desaturase from an umbelliferous plant.

The oils produced by soybeans and rapeseed contain relatively high levels of linolenic acid, compared to many other edible oils such as sunflower and corn oil. When the oil is used for cooking this linolenic acid gives off-flavours which linger in the environment leading to so-called room odours. The problem can be overcome by hydrogenation of the linolenic acid, but this adds significantly to the cost. Consequently breeding programmes have been aimed at reducing the linolenic acid content while retaining the nutritionally desirable linoleic acid content, and of course maintaining favourable agronomic traits (Rattray, 1984). The enzymes responsible for the synthesis of the linolenic acid from linoleic acid are the desaturases associated with the endoplasmic reticulum. The down-regulation of specific desaturase activities would effect an improvement in the quality of soybean and rapeseed oils. Unfortunately these membrane-bound enzymes have proved to be more difficult to purify than the soluble enzymes, and they remain poorly characterised.

The highest price among the main commercial plant fats and oils is obtained for the fat (cocoa butter) from the beans of the cacao tree. The most important feature is its characteristic melting behaviour which makes it sought after in chocolate and confectionery manufacture. The major component of cocoa butter is triacylglycerol of the types *sn*-1,3-saturated, *sn*-2-unsaturated, particularly 1-palmitoyl, 2-oleoyl, 3-stearoylglycerol and 1,3-distearoyl, 2-oleoylglycerol. The enzymes in the developing cacao embryo responsible for the specific arrangement of acyl groups on the glycerol backbone have not yet been identified, but when they are, their transfer to plants like the high-stearate varieties of soybean would provide a new source of cocoa butter (Knauf, 1987).

In principle genetic manipulation could be applied to increase the overall yield of oilseed crops (Knauf, 1987). This would involve the identification of the rate-limiting biochemical step(s) and the relief of that limitation by up-regulating the limiting enzyme. It is assumed that the availability of glycerol is not a limiting factor, because glycerol is readily synthesised from the intermediates of the glycolytic breakdown of photosynthetically derived assimilate. Instead attention is focused on the biosynthesis of the fatty acids. The enzymes that have been identified from *in vitro* studies to be potentially rate-limiting are acetyl-CoA carboxylase, ACP acetyl transferase and ACP malonyl transferase. In particular, acetyl-CoA carboxylase activity has been shown to follow the rate of fatty acid synthesis in rapeseed. However, a definitive identification of rate-limiting enzymes can be made most effectively only when recombinant DNA techniques have provided us with plants containing graded doses of the enzymes proposed to be rate-limiting.

The foregoing has been concerned with some biochemical aspects of the improvement of the current major oil crop species. However, the biotechnological potential for the exploitation of plant oils appears to extend beyond this small group. Between 250 and 300 different fatty acids have been

Table 6.7 Novel oilseed crops and their specialised products.

Species	Fatty acid of interest
Crambe	Erucic acid C22:1
Cuphea	Medium-chain fatty acids especially lauric C12:0
Cape marigold	Dimorphecolic acid C18:2, hydroxy
Milkweed	Vernolic acid C18:1 epoxy
Lesquerella	Lesquerelic acid C20:1, hydroxy
Meadowfoam	Unsaturated very long chain fatty acids, especially 20:1$\triangle$5

identified in plant oils, with chain lengths extending from C8 to C30. In order to meet the industrial demands for oils containing certain of these unusual fatty acids, species that hitherto have not been cultivated on any scale are now being considered as novel oilseed crops (Table 6.7). However, the introduction of an undomesticated species as a new crop requires a long period of breeding in which the plant is altered to meet the requirements of modern agriculture. This delay could be by-passed by transferring the specific biosynthetic ability from the undomesticated species to one of the traditional oilseed crops. Moreover, the techniques of molecular biology now available allow, at least in principle, the magnification of gene expression so that a particular acid of value could become the major component of the oil produced.

FURTHER READING

Anon (1986). Developing new commercial crops, *J. Am. Oil Chem. Soc.* **65**, 6–20.

Anon (1989). Applications of genetically modified oils, *J. Am. Oil Chem. Soc.* **66**, 1058–1061.

Appelqvist, L-A. (1989). The chemical nature of vegetable oils in *Oil Crops of the World*, Eds Röbbelen, G., Downey, R.K. & Ashri, A., London, McGraw-Hill, pp. 22–37.

Battey, J.F., Schmid, K.M. & Ohlrogge, J.B. (1989). Genetic engineering for plant oils: potential and limitations, *Trends in Biotechnol.* **7**, 122–126.

Gunstone, F.D., Harwood, J.L. & Padley, F.B. (1986). *The Lipid Handbook*, London, Chapman and Hall.

Harwood, J.L. (1980). Plant acyl lipids: structure, distribution, and analysis, in *The Biochemistry of Plants. Volume 4 Lipids: Structure and Function*, Ed Stumpf, P.K., New York, Academic Press, pp. 2–55.

Harwood, J.L. (1988). Fatty acid metabolism, *Ann. Rev. Plant Physiol. Plant Mol. Biol.* **39**, 101–138.

Hatje, G. (1989). World importance of oil crops and their products, in *Oil Crops of the World*, Eds Röbbelen, G., Downie, R.K. & Ashri, A., London, McGraw-Hill, pp. 1–21.

Jaworski, J.G. (1987). Biosynthesis of monoenoic and polyenoic fatty acids, in *The Biochemistry of Plants, Volume 9 Lipids: Structure and Function*, Ed Stumpf, P.K., New York, Academic Press, pp. 159–174.

Knauf. V.C. (1987). The application of genetic engineering to oilseed crops, *Trends in Biotechnol.* **5**, 40–46.

Ohlrogge, J.B. (1987). Biochemistry of plant acyl carrier proteins, in *The Biochemistry of Plants. Volume 9 Lipids: Structure and Function*, Ed Stumpf, R.K., New York, Academic Press, pp. 137–157.

Pryde, E.H. & Rothfus, J.A. (1989). Industrial and nonfood uses of vegetable oils, in *Oil Crops of the World*, Eds Röbbelen, G., Downey, R.K. & Ashri, A., London, McGraw-Hill, pp. 87–117.

Quick, G.R. (1989). Oilseeds as energy crops, in *Oil Crops of the World*, Eds Röbbelen, R.K., Downey, R.K. & Ashri, A., London, McGraw-Hill, pp. 118–131.

Rattray, J.B.M. (1984). Biotechnology and the fats and oils industry—an overview, *J. Am. Oil Chem. Soc.* **61**, 1701–1712.

Stumpf, P.K. (1987). The biosynthesis of saturated fatty acids, in *The Biochemistry of Plants. Volume 9 Lipids: Structure and Function*, Ed Stumpf, P.K., New York, Academic Press, pp. 121–136.

Stumpf, P.K. (1989). Biosynthesis of fatty acids in higher plants, in *Oil Crops of the World*, Eds Röbbelen, G., Downey, R.K. & Ashri, A., London, McGraw-Hill, pp. 38–62.

Stymne, S. & Stobart, A.K. (1987). Triacylglycerol biosynthesis, in *The Biochemistry of Plants. Volume 9 Lipids: Structure and Function*, Ed Stumpf, P.K., New York, Academic Press, pp. 175–214.

Vles, R.O. & Gottenbos, J.J. (1989). Nutritional characteristics and food uses of vegetable oils, in *Oil Crops of the World*, Eds Röbbelen, G., Downey, R.K. & Ashri, A., London, McGraw-Hill, pp. 63–86.

ADDITIONAL REFERENCES

Hellyer, A., Bambridge, H.E. & Slabas, A.R. (1986). Plant acetyl-CoA carboxylase, *Trans. Biochem. Soc.* 14, 565–568.

Murphy, D.J., Cummins, I. & Kang, A.S. (1989). Synthesis of the major oil-body membrane protein in developing rapeseed *(Brassica napus)* embryos, *Biochem. J.* 258, 285–293.

Safford, R., Windust, H.C., Lucas, C., De Silva, J., James, C.M., Hellyer, A., Smith, C.G., Slabas, A.R. & Hughes, S.G. (1988). Plastid-encoded seed acyl-carrier protein of *Brassica napus* is encoded by a distinct, nuclear multigene family, *Eur. J. Biochem.* 174, 287–295.

Shimakata, T. & Stumpf, P.K. (1983). The purification and function of acetyl coenzyme A:acyl carrier protein transacylase. *J. Biol. Chem.* 258, 3592–3598.

Stobart, A.K., Stymne, S. & Höglund, S. (1986). Safflower microsomes catalyse oil accumulation *in vitro*: a model system, *Planta* 169, 33–37.

Stymne, S. & Stobart, A.K. (1984). The biosynthesis of triacylglycerols in microsomal preparations of developing cotyledons of sunflower (*Helianthus annuus* L.), *Biochem. J.* 220, 481–488.

Chapter 7

Rubber

INTRODUCTION

In many respects rubber is an extraordinary crop product. A hydrocarbon polymer constructed of isoprene units, natural rubber is a secondary metabolite for which the producing plant has no obvious use. Produced by many species, there is only one source for its commercial production: the secondary phloem of the tropical rubber tree, from which the rubber is tapped over a period of many years. Yet, despite this unlikely biological background, rubber has become an essential raw material for modern industry, a major commodity of world trade, and an important element in the economy of producing countries like Malaysia and Indonesia. The annual production of natural rubber is about 4 million tonnes, over half of which is consumed in the manufacture of motor vehicle tyres.

In the context of the biotechnological exploitation of plants, rubber production is significant on at least two counts. First, it provides an example of how a plant product with unique chemico-physical properties, efficiently produced, can maintain a large market, even in the face of competition from oil-based synthetic substitutes. For the past 40 years natural rubber has been fighting a continuous battle with synthetic rubbers, but nevertheless it retains about 30 per cent of the total market for rubber. Second, as a renewable form of pure hydrocarbon, the future role of rubber may be even greater than at present, especially if the commercial base is widened to include a greater range of rubber-producing species.

STRUCTURE OF RUBBER

Molecules of rubber are made up of long, unbranched chains of 1,4-isoprene units, with a M_r of 10^5–10^7. Formerly, rubber was believed to be made up

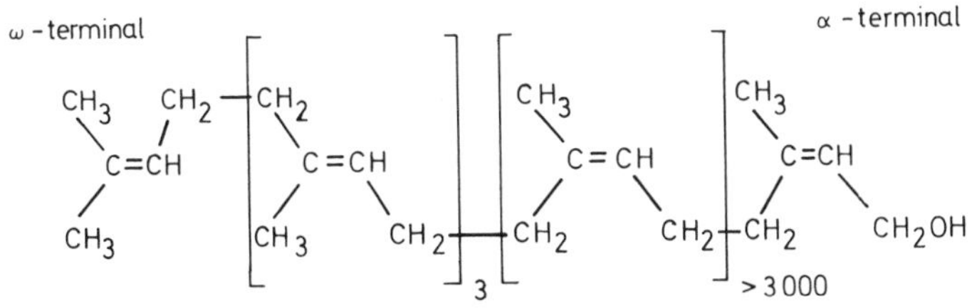

Fig. 7.1 Structure of the rubber molecule.

entirely of *cis*-polyisoprene, but it is now known that each chain is initiated by three or four isoprene units in which the double bond has the *trans*-configuration (Figure 7.1).

Unstretched, the long, flexible rubber molecules take up a tangled configuration. But when the rubber is stretched, the disordered molecules become aligned, and then when the tension is released, they regain their original random configuration. This elasticity of rubber depends on the predominantly *cis*- configuration of the polymer. The chemically related product, gutta, is a *trans*-polyisoprene, and is partly crystalline in the unstretched condition, so that it is harder than rubber, and can be stretched only on warming. For many applications natural rubber if strengthened by a process called vulcanisation, which was discovered by Goodyear in 1839. In this process the *cis*-isoprene polymers of rubber are crosslinked by sulphur and other reagents to give a three-dimensional network which strengthens the rubber while retaining its unique elasticity.

There is no satisfactory explanation for the function of rubber in plants such as the rubber tree, although a number of proposals have been made. The rubber particles help latex to coagulate on exposure to air, and consequently it has been suggested that latex helps to protect against a mechanical injury. The rubber once deposited, is not metabolised and hence it does not serve as a reserve of energy or reduced carbon. It has been suggested that rubber formation is effectively a fermentation process, serving to consume excess reducing equivalents, but supporting evidence is lacking. Similarly, the formation of rubber might function to mop up photosynthate produced in excess of demand from growth: rubber production in the rubber tree does diminish during seasons of partial leaf loss; and in the Guayule shrub, rubber accumulates especially during winter when photosynthesis continues in the absence of growth. Whatever the function of rubber in the producing plants, however, varieties which lack its production seem to suffer no disadvantage (Backhaus, 1985).

RUBBER PRODUCTION

Rubber is formed in about 2000 species of plants belonging to seven families, but generally the amounts are small and the rubber is of poor quality because the polymeric chains are too short, and the entire commercial production comes from the rubber tree. However, plants which have been minor sources in the past are now being re-examined as potential rubber-yielding crops for regions outside the humid tropics. In the rubber euphorbia, the rubber, produced in latex, is of inferior quality, but the plant is resistant both to drought and to salinity, and therefore with improvement it offers a potential crop for otherwise unproductive marginal lands. In the Guayule shrub the rubber, with a M_r of 2–20 × 10⁵, is of excellent quality. It is stored in the parenchyma, not in latex, so to harvest the rubber the shrub is cut to the ground. The plant regrows to provide further harvests. The Guayule flourishes in arid conditions and is amenable to mechanical harvesting. It therefore may offer in the future some competition to the rubber tree as a source of natural rubber. However, the enormous advantage of the rubber tree is that a good yield of high quality rubber can be obtained continuously by tapping the bark of a mature stand of trees.

In the rubber tree the rubber is formed as a suspension of particles in the milky latex of the secondary phloem. The latex is contained within latex vessels or laticifers, which form a jointed, interweaving network of tubes interspersed among the phloem vessels and the thin plates of medullary rays of the secondary phloem (Figure 7.2).

There are many different techniques for tapping the latex, but in one of the most common, tapping is initiated by cutting a spiral groove in the bark for

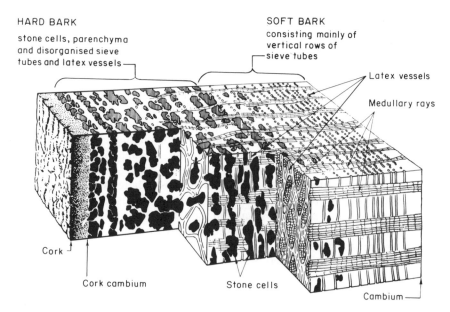

HARD BARK
stone cells, parenchyma
and disorganised sieve
tubes and latex vessels

SOFT BARK
consisting mainly of
vertical rows of
sieve tubes

Latex vessels

Medullary rays

Cork

Cork cambium   Stone cells

Cambium

Fig. 7.2   Bark anatomy of the rubber tree (courtesy of the Malaysian Rubber Producers' Research Association). Note that the sieve tubes which make up the bulk of the soft bark, and parenchyma in both soft and hard bark, are not shown in the diagram. The black areas are clusters of stone cells.

half the circumference of the trunk (Figure 7.3). The groove cuts almost into the cambium so as to sever the latex vessels in the so-called soft bark. Tapping is then carried out on alternate days by shaving a thin layer of bark from the lower edge of the groove to re-open the plugged latex vessels of the bark and release the latex upward into the new cut (Paardekooper, 1989). The latex is collected by allowing it to run into a small cup attached to the tree. With successive tappings the groove moves down the trunk of the tree, and as it does so the bark above is regenerated. After about 5 years the base of one side of the trunk is reached and the other side is then exploited in the same way. By the time the base of the second side of the trunk has been reached the bark on the original side has regenerated by the action of the secondary cambium, and tapping can be restarted on the first side. In this way 20 years

Fig. 7.3 Tapping the bark of the rubber tree (courtesy of the Malaysian Rubber Producers' Research Association).

of continuous tapping can be obtained with virgin bark and the bark of one renewal.

As for any crop the factors that affect the productivity of the rubber trees are environmental, cultural and genetic. Included in the genetic factors are the number of latex vessels, and the thickness of the secondary phloem. But the cells that surround the latex vessels also provide assimilates to regenerate the latex withdrawn on tapping, and these cells are therefore also important in determining productivity. Tapping yields in a year, per tree, about 6–40 1 of latex containing about 30 per cent rubber. This translates into a maximum annual yield of about 9000 kg of dry rubber per hectare from a mature stand of rubber trees (Paardekooper, 1989).

LATEX

The latex vessels develop by the dissolution of the cross walls of cells which possess a nucleus and a cytoplasm containing the usual organelles: ribosomes, endoplasmic reticulum, Golgi apparatus, mitochondria and proplastids. As the vessels mature the lumen becomes increasingly occupied by numerous rubber particles so that the original cytoplasm and its organelles become confined to the sides of the vessels.

The rubber particles range in size from 0.005 to 3 μm (Figure 7.4), each being surrounded by an osmiophilic membrane containing protein and phospholipid. There are two other kinds of particle found in latex (Figure 7.4). The more numerous are the lutoids. These are fragile, osmotically sensitive vesicles containing divalent cations dissolved in an acid sap. When they burst the acidity of the sap and the divalent cations are released which causes the rubber particles to flocculate (Chrestin *et al.*, 1984b). This flocculating effect helps to stop the flow of latex after tapping. The lutoids burst when the unsaturated lipids of the membrane are attacked by superoxide anions (O_2^-), hydroxyl radicals (OH·) and hydrogen peroxide (H_2O_2). These toxic species in turn are generated by a NAD(P)H oxidase at the lutoid membrane. Activity of this enzyme is stimulated by Fe^{3+} and Cu^{2+}, which are normally sequestered within the lutoids but released when the lutoids burst. The cations released from the burst lutoids neutralise the negative charge at the surface of the rubber particles, and thus cause the rubber particles to flocculate. The released cations also stimulate the NAD(P)H oxidase activity, and therefore their release sets in train an autocatalytic process of lutoid destruction. Rubber trees in which the flow of latex is restricted are said to be suffering from 'dry-bark disease'. Compared to the latex of normal trees, the latex of these trees has fewer and more fragile lutoids, with a higher lutoid NAD(P)H oxidase activity (Chrestin *et al.*, 1984a).

The lutoids also play an important role in maintaining optimal conditions for the regeneration of latex after tapping. The productivity of rubber trees measured in terms of the latex produced per tapping is positively correlated with the pH of the latex cytosol: the more acid the cytosol, the less rubber

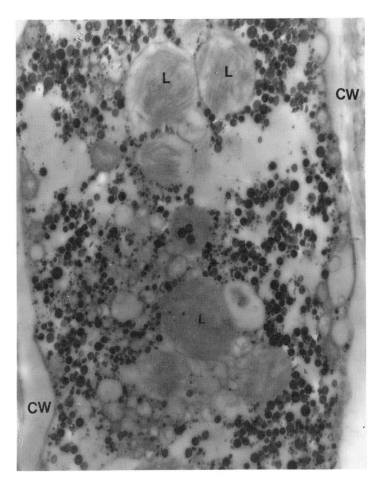

Fig. 7.4 Electron micrograph of latex vessel showing cell wall (CW), lutoids (L) and numerous darkly staining rubber particles (× 9000) (courtesy of the Malaysian Rubber Producers' Research Association).

that is regenerated between tappings. The lutoid tonoplast contains two H^+-translocating enzymes: an ATPase and an NADH-cytochrome c oxidoreductase. The ATPase pumps protons into the lutoid while the NADH-cytochrome c oxidoreductase pumps protons out of the lutoid. When the cytosolic pH falls to the lower end of the physiological range (pH 6.5) the ATPase is at its pH optimum, and is more active than the NADH-cytochrome c oxidoreductase, which has a pH optimum at about pH 7.6 (Chrestin *et al.*, 1984b). Thus with an acid cytosol there would be a net movement of protons into the lutoids from the cytosol. Conversely if the cytosolic pH were to become too high, the NADH-cytochrome c oxidoreductase would tend to acidify the cytosol. In this way it is proposed that the lutoid tonoplast acts as a 'pH-stat' helping to maintain an optimum cytosolic pH for rubber synthesis (Chrestin *et al.*, 1984b).

Besides the lutoids, the other characteristic feature of the latex is the Frey–Wyssling particle which consists largely of lipid coloured orange/yellow by carotenoid. Two or three of these particles clustered together and surrounded by a double membrane form the Frey–Wyssling complex. The inner of the two surrounding membranes is invaginated and folded. The function of the Grey–Wyssling complex is unknown, but the double membrane and the presence of carotenoid suggest that it is derived from a plastid.

BIOSYNTHESIS

The biosynthetic pathway of rubber can be divided into three stages (Figure 7.5):

1. The generation of acetyl-CoA
2. The conversion of acetyl-CoA to isopentenyl diphosphate
3. The polymerisation of isopentenyl diphosphate units to rubber.

All three stages can be demonstrated in tapped latex freshly drawn from the rubber tree bark. In general, the evidence in support of this pathway has been obtained from experiments in which the suspected intermediate, in a radioactively labelled form, has been fed to the tapped latex, and by assaying for the suspected enzyme activity. The first two stages are located in the soluble phase of the latex, called the serum, while the third stage occurs only at the surface of the rubber particles (Figure 7.5).

The serum, contains all the enzymes of the glycolytic sequence. It has generally been assumed that the acetyl-CoA was derived from pyruvate produced by the glycolytic breakdown of sugars. However, ^{14}C-labelled pyruvate and its carbohydrate precursors are poorly incorporated into rubber. One possible explanation for the poor incorporation is that the pyruvate dehydrogenase responsible for the conversion of pyruvate to acetyl-CoA is confined to the latex mitochondria (as it is in other plant tissues) and therefore the enzyme is poorly represented in the tapped latex, because most of the latex mitochondria are retained in a parietal layer of the latex vessels. However, even if the acetyl-CoA for rubber synthesis were generated by a mitochondrially located pyruvate dehydrogenase it would normally be utilised within the mitochondria, while it is known that isopentenyl diphosphate synthesis from acetyl-CoA occurs in the latex serum. Thus there may be alternative sources of acetyl-CoA for rubber synthesis such as β-oxidation of fatty acids or the metabolism of branched chain amino acids.

The conversion of acetyl-CoA to isopentenyl diphosphate follows the terpenoid pathway used by plants for the synthesis of sterols and carotenoids (Figure 7.6). Two successive condensation reactions convert the acetyl-CoA first to acetoacetyl-CoA, then to 3-hydroxy 3-methylglutaryl-CoA. This six-carbon intermediate then loses its CoA and is reduced to mevalonic acid using NADPH, which is most likely generated by the pentose phosphate pathway. The reductase in latex is particulate, and probably located in the endoplasmic reticulum. In the synthesis of cholesterol in animals, 3-hydroxy 3-methylglutaryl-CoA reductase is important as a rate-limiting step, and may be so in rubber synthesis. The conversion of mevalonic acid to isopentenyl diphosphate occurs via the formation of mevalonate 5' phosphate and mevalonate 5' diphosphate (Figure 7.6).

As a prelude to polymerisation there is an isomerisation of isopentenyl diphosphate to dimethylallyl diphosphate catalysed by isopentenyl diphosphate isomerase (Figure 7.5). Polymerisation then takes place by the successive

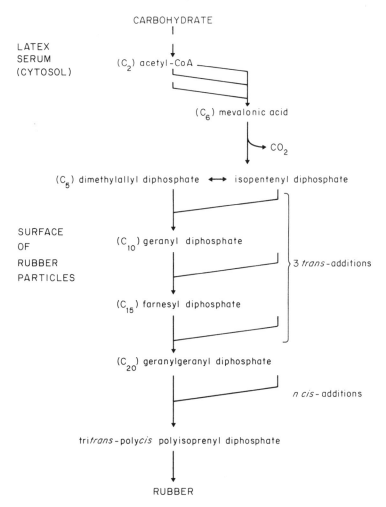

Fig. 7.5 Outline of rubber biosynthesis in the latex of the rubber tree.

Table 7.1 Enzymes responsible for the synthesis of rubber from acetyl-CoA.

1. Acetyl-CoA acetyltransferase
   Acetyl-CoA:acetyl-CoA C-acetyltransferase $\qquad$ EC 2.3.1.9

   acetyl-CoA + acetyl-CoA → CoA + acetoacetyl-CoA

2. Hydroxymethylglutaryl-CoA synthase
   (S)-3-hydroxy-3-methylglutaryl-CoA acetoacetyl-CoA-lyase
   (CoA acetylating) $\qquad$ EC 4.1.3.5

   acetoacetyl-CoA + acetyl-CoA + H_2O ⟷ 3-hydroxy-3-methylglutaryl-CoA + CoA

3. Hydroxymethylglutaryl-CoA reductase
   (S)-mevalonate:$NADP^+$ oxidoreductase $\qquad$ EC 1.1.1.34

   3-hydroxy-3-methylglutaryl-CoA + 2 NADPH → mevalonate + CoA + 2 NADP

4. Mevalonate kinase
   ATP:(R)-mevalonate 5′-phosphotransferase $\qquad$ EC 2.7.1.36

   ATP + (R)-mevalonate → ADP + (R)-5′-phosphomevalonate

5. Phosphomevalonate kinase
   ATP:(R)-5′phosphomevalonate phosphotransferase $\qquad$ EC 2.7.4.2

   ATP + (R)-5′-phosphomevalonate → ADP + (R)-5′-diphosphomevalonate

6. Diphosphomevalonate decarboxylase
   ATP:(R)-5′-diphosphomevalonate carboxylyase (dehydrating) $\qquad$ EC 4.1.1.33

   ATP + (R)-5′-diphosphomevalonate → ADP + P_i + isopentenyl diphosphate + CO_2

7. Isopentenyl diphosphate △-isomerase
   isopentenyl diphosphate $\triangle^3$-$\triangle^2$-isomerase $\qquad$ EC 5.3.3.2

   isopentenyl diphosphate ⟷ dimethylallyl diphosphate

8. Dimethylallyl-*trans*-transferase
   dimethylallyl diphosphate:isopentenyl diphosphate
   dimethylallyl-*trans*-transferase $\qquad$ EC 2.5.1.1

   dimethylallyl diphosphate + isopentenyl diphosphate → PP_i + geranyl diphosphate

9. Geranyl-*trans*-transferase
   geranyl diphosphate:isopentenyl diphosphate
   geranyl-*trans*-transferase $\qquad$ EC 2.5.1.10

   geranyl diphosphate + isopentenyl diphosphate → PP_i + *trans, trans*-farnesyl diphosphate

10. Rubber transferase
    poly-*cis*-polyprenyl diphosphate:isopentenyl diphosphate
    polyprenyl-*cis*-transferase $\qquad$ EC 2.5.1.20

    poly-*cis*-polyprenyl diphosphate + isopentenyl diphosphate → PP_i + a poly-*cis*-polyprenyl diphosphate longer by one C_5 unit additions of isopentenyl diphosphate to dimethylallyl diphosphate to form, in order, geranyl diphosphate (C_{10}), farnesyl diphosphate (C_{15}), and geranylgeranyl diphosphate (C_{20}), as shown in Figure 7.7. It is now clear that this initial linking of isoprenyl units occurs with the *trans*-configuration. By contrast, the subsequent repeated addition of isoprenyl units occurs in the *cis*-configuration to give rise to the bulk of the 30 000 isoprenyl units of the final rubber molecules. This polymerisation occurs on the surface of the rubber particles,

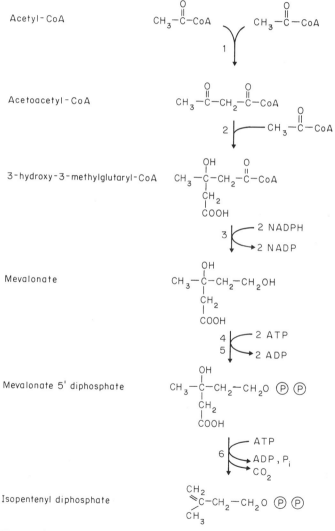

Fig. 7.6 Formation of isopentenyl diphosphate from acetyl-CoA in rubber biosynthesis. Figures refer to the enzymes shown in Table 7.1.

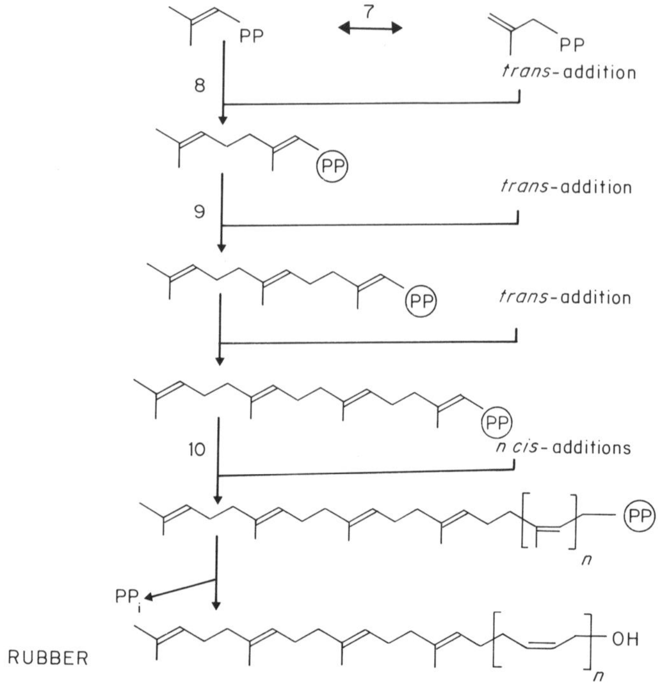

Fig. 7.7 Polymerisation of isopentenyl diphosphate to form rubber. Figures refer to the enzymes shown in Table 7.1.

and is catalysed by an enzyme referred to as the rubber transferase. Thus the biosynthesis of rubber from isopentenyl diphosphate requires enzymes capable of catalysing the isomerisation of isopentenyl diphosphate to dimethylallyl diphosphate, then at least two prenyl transferases to catalyse the transfer of the three or four isoprene units in the *trans-* configuration to generate farnesyl and geranylgeranyldiphosphate, and prenyl transferases to catalyse the transfer of the large number of isoprene units in the *cis-* configuration, the rubber transferase. Termination of polymerisation is important for the quality of the rubber, since in principle the length of the rubber molecules will be determined by the timing of this reaction, but it remains poorly understood. Possibly a phosphatase dephosphorylates the rubber diphosphate after it has reached a given length.

Since each rubber particle contains of the order of 10^6 monomers, particle initiation is a rare occurrence compared to polymerisation, and this makes the study of chain initiation more difficult than that of chain elongation. Possibly the rubber particles originate as molecules of geranylgeranyldiphosphate attached to the transferase enzyme in the latex serum (Archer & Audley, 1987). Repeated additions of isopentenyl diphosphate and more rubber molecules would then result in the formation of the spherical rubber particle surrounded by enzymes and other molecules constituting the osmiophilic film that can be seen in the electron microscope.

BIOTECHNOLOGICAL DEVELOPMENT

It is now recognised that rubber is not the unique plant product that it was once thought to be. Since the terminal *trans*-isoprene units have been identified as a feature of rubber (Tanaka, 1989), it has become apparent that rubber may be viewed as a very high homologue of a widely distributed family of oligo-*trans*-poly-*cis*-isoprenols, where a short length of *trans*-linked isoprene units terminates the predominantly *cis*-linked isoprene chain. For example, the ficaprenols found in the India rubber tree consist of five to nine *cis*-isoprene units terminated by three *trans*- units. The function of none of these shorter oligo-*trans*-poly-*cis*-isoprenols is known. Presumably they originated in evolution by the mutation of the gene for a prenyltransferase in the terpenoid pathway, such as the geranyldiphosphate synthetase enzyme, so that the mutated enzyme then catalysed a transfer of isoprenyl units to the growing chain in a *cis*-configuration and in an unregulated manner. In the case of *Hevea* rubber, the chain grew to an unprecedented length, elongation being catalysed by an enzyme which, like the Sorcerer's Apprentice, became virtually unstoppable. Alternatively, the synthesis of *Hevea* rubber may have originated from the failure of—or a long delay in the action of—the mechanism that normally terminates the *cis*-elongation of the shorter oligo-*trans*-poly-*cis*-isoprenol chains.When the genes responsible for these reactions have been sequenced then a clearer picture of the evolutionary origins of rubber production will emerge.

An important point to arise from these evoluntionary considerations is that synthesis of a high quality rubber seems to depend on the presence of a single enzyme, rubber transferase. If rubber production depends solely on the action of rubber transferase superimposed on the normal terpenoid pathway, then transfer of the gene for rubber transferase to other plants would spread the ability to synthesise rubber to an unlimited range of plants—all green plants are able to synthesise the linear isoprenoid that constitutes the phytyl side chain of chlorophyll. However, the cellular organisation of rubber synthesis is not yet fully understood, and this may limit the accumulation of sufficiently large amounts for commercial exploitation. Moreover the quantities accumulated may be limited by the supply of isopentenyl diphosphate, acetyl-CoA, ATP and NADPH.

A high M_r is essential for the commercial utilisation of rubber. Plants such as the sunflower and goldenrod produce small amounts of a rubber which, with a M_r of less than 5×10^4, is of inferior quality. The rubber of *Hevea* and Guayule contains chains which are ten times that length. In *Hevea* the mean M_r is genetically controlled, and differs significantly between different clones. When we have identified the enzymes that are responsible for chain termination and thus determine chain length, the way will become open for the genetic transformation of species presently producing inferior rubber so that they are capable of producing a commercially useful product.

FURTHER READING

Archer, B.L. (1980). Polyisoprene, in *Encyclopedia of Plant Physiology, New Series. Volume 8 Secondary Plant Products*, Eds Bell, E.A. & Charlwood, B.V., Berlin, Springer-Verlag, pp. 309–327.
Archer, B.L. & Audley, B.G. (1987). New aspects of rubber biosynthesis, *Bot. J. Linnean Soc.* **94**, 181–196.
Backhaus, R.A. (1985). Rubber formation in plants—a minireview, *Israel J. Bot.* **34**, 283–293.
Baulkwill, W.J. (1989). The history of natural rubber production, in *Rubber*, Eds Webster, C.C. & Baulkwill, W.J., Harlow, Longman pp. 1–56.
D'Auzac, J., Jacob, J.-L. & Chrestin, H. (Eds.) (1989). *Physiology of Rubber Tree Latex*. Boca Raton, Florida, CRC Press.
Morris, J.E. (1989). Processing and marketing in *Rubber*, Eds Webster, C.C. & Baulkwill, W.J., Harlow, Longman, pp. 459–498.
Paardekooper, E.C. (1989). Exploitation of the rubber tree, in *Rubber*, Eds Webster, C.C. & Baulkwill, W.J., Harlow, Longman, pp. 349–414.
Paterson-Jones, J.C., Gilliland, M.G. & Van Staden, J. (1990). The biosynthesis of natural rubber, *J. Plant Physiol.* **136**, 257–263.
Webster, C.C. & Paardekooper, E.C. (1989). The botany of the rubber tree, in *Rubber*, Eds Webster, C.C. & Baulkwill, W.J., Harlow, Longman, pp. 57–84.

ADDITIONAL REFERENCES

Chrestin, H., Bangratz, J., D'Auzac, J. & Jacob, J.L. (1984a). Role of the lutoidic tonoplast in the senescence and degeneration of the laticifers of *Hevea*, *Z. Pflanzenphysiol.* **114**, 261–268.
Chrestin, H., Gidrol, X., Marin, B., Jacob, J.L. & D'Auzac, J. (1984b). Role of the tonoplast in the control of the cytosolic homeostasis within the laticiferous cells of *Hevea*, *Z. Pflanzenphysiol.* **114**, 269–277.
Gomez, J.B. & Moir, G.F.J. (1979). The ultracytology of latex vessels in *Hevea brasiliensis, Monograph No 4. Malaysia Rubber Research Development Board*, Kuala Lumpur.
Tanaka, Y. (1989). Structure and biosynthesis mechanism of natural polyisoprene, *Progr. Polymer Sci.* **14**, 339–371.

Chapter 8

Protein

INTRODUCTION

The protein in our diet and the protein we feed to our livestock comes largely from the seeds of cereals and legumes. Compared to the vegetative parts of a plant, these seeds hold a concentrated store of protein. As a proportion of their dry weight, leaves contain 3 to 5 per cent protein, cereal seeds 8 to 15 per cent, and legumes 20 to 30 per cent, which can rise to 40 per cent in the case of soybean. The protein in tuber crops such as the potato, which generally contains about 5 per cent protein, can also make an important contribution to the protein we eat. However, the dry state and low metabolic activity of seeds allow them to be easily stored and widely traded. Among the cereals wheat is the most widely grown with a world production of about 500 million tonnes per year; and among the legumes soybean leads with a production of about 100 million tonnes per year (Figure 6.1). The production figures for the other legume protein sources are much lower (Figure 8.1). A further important source of protein is the cake which remains when oil has been extracted from those seeds that are grown principally for their oil (Chapter 6) such as oilseed rape and sunflower (Figure 6.1). Since it is in seeds rather than in the vegetative parts of the plant that protein is accumulated in the stable and concentrated form that makes it an important crop product, it is specifically seed protein that we will be concerned with in this chapter.

Seeds do not accumulate protein for our benefit, but to serve as a reserve of nitrogen (and of sulphur and carbon) for the seedling during germination and in the early stages of growth before an adequate supply of mineral nutrients has been obtained from the soil. Thus more than half the proteins of a seed are storage proteins with no known catalytic function. These storage proteins

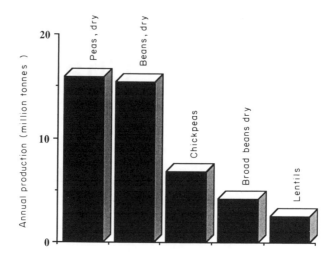

Fig. 8.1 World production of cereals and pulses (averages for the years 1987–89); (from FAO Production Yearbook, 1989).

have evolved to fulfil a particular role: to be laid down as a nitrogen-rich store at one stage of growth in order for the nitrogen to be mobilised at a subsequent stage. This may appear to be a relatively undemanding role for a molecule such as a protein which is capable of great subtlety in its primary sequence and three-dimensional structure. It can also appear to be a role that is undemanding when compared to the precisely regulated catalytic function of an enzyme protein. However, such expectations may well be mistaken, since analysis of the molecular structures of storage proteins of a variety of types and from different species reveals them to be highly conserved. This suggests that there are restraints on structural variation that have been retained over long periods of time.

What demands are made on storage proteins? They must contain a rela-tively high ratio of N:C in a compact structure; react appropriately to desiccation when the seed dries and to rehydration when the seed germinates; maintain a stable structure in the dry state; finally, they must be readily broken down by protease action during seed germination. In a sense these are the criteria for seed protein quality that are applied by the plant. Protein quality means something quite different to the user of the crop. In wheat destined for breadmaking, particular protein fractions need to possess those physical qualities that will allow the bread to rise—a consideration of com-plete indifference to the seed. In cereals grown for feed, the amino acid composition is often seriously imbalanced compared with the nutritional requirements of the livestock, with lysine being deficient and the amide glutamine predominating. When the cereal grain germinates the storage protein serves as a source of nitrogen rather than of amino acids, since the growing regions of the young plant are able to synthesise *de novo* the 20

Table 8.1 The amino acids of proteins.

Non-essential	Essential
Glycine (gly)	Threonine (thr)
Serine (ser)	Lysine (lys)
Proline (pro)	Arginine (arg)
Cysteine (cys)	Histidine (his)
Aspartate (asp)	Valine (val)
Glutamate (glu)	Leucine (leu)
Asparagine (asp)	Isoleucine (ile)
Glutamine (gln)	Phenylalanine (phe)
Alanine (ala)	Tryptophan (trp)
Tyrosine (tyr)	Methionine (met)

protein amino acids. However, we and our pigs and poultry require in our diet about 10 particular amino acids, the essential amino acids. When these essential amino acids are deficient they limit the utilisation of the dietary protein for synthesis and instead it is consumed as an energy source, and the excess nitrogen excreted. Thus man and the seed-producing plant make different technical and nutritional demands on the storage protein.

The eventual aim of much of the research into the biosynthesis and molecular structure of storage proteins is to acquire the ability to manipulate these proteins so that they more effectively meet our demands, while still fulfilling the plant's requirements and without jeopardising the productivity of the crop. However, in order to understand how storage proteins are made, and how their structure and properties are related to their function, it is essential to know something of the different types of storage protein accumulated by the major crops.

NATURE OF PLANT STORAGE PROTEINS

Plant proteins are still classified into the groups first distinguished by Osborne in the early years of this century. The Osborne Classification of proteins is based on the extractibility of the proteins in a sequential series of solvents (Table 8.2).

Proteins that are soluble in water are classed as albumins. Since these proteins have a predominantly metabolic rather than storage function, they do not enter into the present description. The true storage proteins are virtually insoluble in water. The globulins are characterised by their solubility in dilute salt solutions. They are widely distributed among plant families, being the major form of storage protein in the dicotyledonous plants, including the legumes, but in the cereals they are largely replaced by the glutelins and prolamins. Notable exceptions to this rule are oats, where the major protein is a globulin. The prolamins are insoluble in water and salt solutions,

Table 8.2 Solubility and distribution of the principal storage proteins of seeds.

Class	Soluble	Insoluble	Distribution
Globulins	Salt solutions	Water	Widely distributed legumes, oilseeds, oats
Prolamins	Aqueous alcohol	Water Salt solutions	Cereal, starchy endosperm
Glutelins	Acid/alkali	Water Salt solutions Aqueous alcohol	Rice

and are extracted into aqueous alcohol. Glutelins are insoluble in the other solvents, and are extracted into acids or alkalis. They constitute the major storage protein of rice.

Proteins need to be brought into solution before they can be studied biochemically. Thus the Osborne solubility classes remain useful operational definitions, even if, as we shall see later, analysis of molecular structure has diminished the significance of differential solubility as a fundamental characteristic. For example, an important component of the protein of wheat dough, the glutenin, is included in the prolamins even though in the native state intermolecular disulphide bonds render it insoluble in aqueous alcohol. When this protein fraction is treated with sulphhydryl-reducing agents, such as mercaptoethanol, the disulphide bonds are broken and the individual subunits are released and become soluble in aqueous alcohol. Moreover, amino acid sequences of proteins from different Osborne classes often show strong homology, which speaks for a closer evolutionary and functional relationship than that suggested by the primary division of the storage proteins into their solubility classes.

The proteins which make up each Osborne solubility class are separable by SDS-PAGE on the basis of their molecular mass, and by IEF on the basis of their charge. This is shown for the prolamins of maize in Figure 8.2. These techniques reveal that a particular fraction is composed of a small number of subclasses of proteins each consisting of many polypeptides which differ slightly in size and charge but which are nevertheless closely related in structure and sequence. The storage proteins are rarely the product of single genes, but are more commonly produced by multigene families.

Globulins

Two quite different classes of globulin have been distinguished, each named after the class-type: the legumins and the vicilins. The legumins are the larger (11S) globulins with a M_r of about 360 000. Each protein consists of six non-identical subunits, and each subunit is made up of an acidic (M_r about 40 000) and a basic polypeptide (M_r about 20 000). The basic and acidic polypeptides

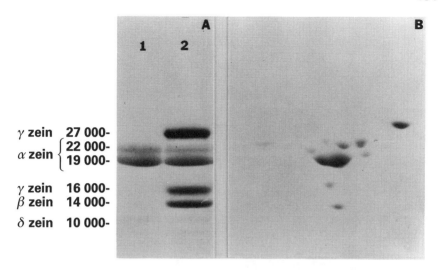

Fig. 8.2 Analysis of the zeins of maize. A. SDS-PAGE of mature endosperm extracted with 70% ethanol (lane 1) or 70% ethanol plus 1% mercapto-ethanol (lane 2). B, SDS-PAGE plus IEF of developing endosperm extracted with 70% ethanol plus 1% mercaptoethanol. (From Larkins *et al.*, 1984. Figure courtesy of Dr B.A. Larkins. Reproduced from *Trends in Biochemical Sciences* by kind permission of Elsevier Trends Journals.)

are linked by a disulphide bond. Figure 8.3 shows that bands representing the acidic and basic subunits of the oat globulin are revealed after the disulphide bonds of the main subunits of M_r 53 000–58 000 are broken by reduction with mercaptoethanol. Proteins from a wide variety of protein-yielding crops are of the legumin-type (Table 8.3).

The smaller (7S) vicilin-type with a M_r of about 200 000 have a more complex subunit structure and there are no linking disulphide bonds. Vicilins are often glycosylated with mannose and glucosamine—for no obvious func-tional purpose, since when glycosylation is specifically inhibited the vicilin is synthesised and deposited normally. Vicilins have mainly been studied in the legumes, notably the phaseolin of French bean and vicilin of pea, but a vicilin-type globulin is also well represented in cottonseed.

When the amino acid sequences of globulins from different sources have been compared, then extensive sequence homology becomes apparent. From these homologies it has been possible to propose an evolutionary develop-ment of the legumins and vicilins from two ancestral types (Borroto & Dure, 1987).

The globulins are made up of amino acids which are nutritionally well balanced in that they are present in similar proportions to the amino acids in animal protein, except for relatively low levels of the sulphur containing amino acids, methionine and cysteine, with the vicilin type being particularly deficient.

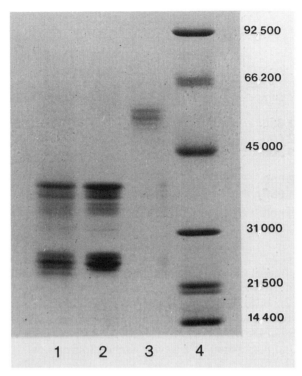

Fig. 8.3 SDS-polyacrylamide gel electrophoresis of oat globulin. Lanes 1 and 2 were obtained after reduction with mercaptoethanol, lane 3 before reduction, and lane 4 shows the M_r markers. (From Brinegar & Peterson, 1982. Figure courtesy of Dr D.M. Peterson. Reproduced from *Archives of Biophysics and Biochemistry* by kind permission of Academic Press.)

Glutelins

In rice the major storage protein can be solubilised with NaOH at pH 12, after the albumins and globulins have been extracted with water and salt solutions. The molecule appears to be made up of three subunits of M_r 38 000, 25 000 and 16 000. The middle-sized component shows significant homology with the small subunit of pea legumin. Thus although there must be sufficient difference in primary and secondary structure to account for the different solubilities, the affinity between the rice glutenin and pea legumin suggests that there may be a closer evolutionary and functional relationship than that suggested by their separation into different Osborne classes.

Prolamins

Prolamins make up about half the storage protein of wheat, barley, maize and sorghum. The term prolamin refers to the predominance of proline and

Table 8.3   The globulin storage proteins of crop plants.

Legumin types (11S)

    Groundnut, peanut
    Soybean (glycinin)
    Lupin
    Pea (legumin)
    Field bean, broad bean
    White mustard
    Sunflower (helianthin)
    Oilseed rape (cruciferin)
    Barley
    Oats (oat globulin)
    French bean

Vicilin types (7S)

    Pea (vicilin)
    French bean (phaseolin)
    Soybean (conglycinin)

glutamine in their amino acid composition. Prolamins generally contain relatively little lysine, so that those cereals with high prolamin levels are correspondingly deficient in lysine, while oats and rice, with about 10 per cent prolamin, have significantly higher levels of lysine (Table 8.4).

Prolamins can be a bewildering class of proteins, consisting as they do of a complicated mixture of a large number of different protein components that vary in size and shape. Nevertheless, there are two common features of the prolamins that help to unify them as a group. First, the molecules can all be divided into regions called domains, which have characteristic amino acid

Table 8.4   The prolamin and lysine contents of seed proteins (from Doll, 1984).

Cereal	Prolamin (Seed N %)	Lysine (Protein %)
Rice	8	3.5
Oats	12	4.2
Barley	40	3.5
Wheat	45	3.1
Maize	50	1.6
Sorghum	60	2.1

Table 8.5 Components of the prolamin fraction of some cereals.

	Barley	Wheat	Rye
HMW prolamins	D hordein	HMW prolamins	
S-poor prolamins	C hordein	ω-gliadin	ω-secalin
S-rich prolamins	B hordein	α-,β-,γ-gliadins	γ-gliadins

compositions. Second, in all the prolamins at least one of these domains is made up of repeated sequences. In many cases it is the dominating influence of these repetitive sequences that gives prolamin proteins their particular three-dimensional structure and amino acid composition. Thus the repetitive sequences can be considered the basis, at the molecular level, of the nutritional and technological qualities of the cereal accumulating that prolamin.

In wheat, barley and rye, which all belong to the sub-family Triticeae, the polypeptides separated on SDS-PAGE have been classed into three major groups: the high molecular weight (HMW) prolamins, the S-poor and the S-rich prolamins (Table 8.5). In all three cereals the bulk of the prolamin (about 80 per cent) is present as the S-rich components, although the S-poor C hordein in barley and the HMW prolamin in wheat can be present in significant amounts.

Of the HMW prolamins, those of bread wheats have been most closely studied because of their relationship with the breadmaking quality of flour. Individually, the HMW protein subunits are soluble in alcohol/water mixtures; it is the presence of disulphide bonds between the chains that causes them to form larger complexes which are insoluble in the alcohol/water mixtures. When they are extracted as an alcohol-insoluble fraction they constitute the glutenin fraction. The characteristic structure of a HMW subunit consists of a N-terminus sequence of 81–104 residues, a long central repetitive domain containing repeated 6-residue, 9-residue and 3-residue sequences, and a C-terminal sequence of 42 residues (Figure 8.4). The cysteine residues which are responsible for the disulphide bonding are confined to the flanking terminal sequences.

The quantitatively most important group of prolamins, the S-rich prolamins, all have a similar general structure which features a N-terminal domain of repeated sequences rich in proline and a C-terminal domain to which virtually all the cysteine residues are confined (Figure 8.4). The high cysteine content allows for a high degree of intra- and intermolecular disulphide bonding, which means that extraction is facilitated by the addition of sulphhydryl-reducing agents to break the disulphide bonds.

The best known of the S-poor prolamins is the C hordein of barley. The molecule is dominated by the central domain of repeated 8-residue units, flanked by short sequences at the N-terminus and C terminus (Figure 8.4). The absence of cysteine in this S-poor prolamin means that the disulphide

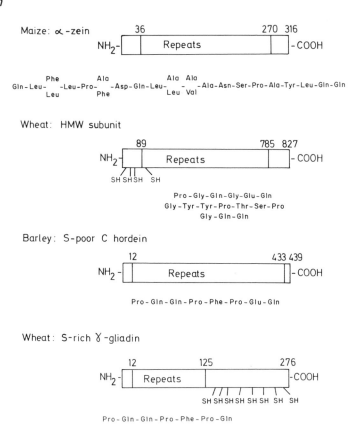

Fig. 8.4 Primary structures of some prolamins showing the amino acids of the repeated sequences (data from Shewry & Tatham, 1990).

bonding of the other prolamins is absent, and the S-poor prolamins are present as monomers.

In maize the prolamins are called zeins. They vary in size from about M_r 10 000 to M_r 27 000, with about 70 per cent of the protein being present in the form of α-zeins (Figure 8.2). These can be resolved by IEF into a large number of differently charged polypeptides, which are encoded for by a correspondingly large family of 75–100 genes. By contrast the quantitatively less important β-, γ- and δ-zeins exhibit little charge heterogeneity. The α-zeins resemble other prolamins in that the major part of the polypeptide is represented by a repetitive domain consisting of a motif of about 20 residues repeated nine times (Figure 8.4). But here the resemblance seems to end, since when the amino acid sequences of the α-zeins are compared with those of other zeins and with the prolamins of other cereals, no homology is apparent.

Protein bodies

Storage protein is accumulated as a dense mass contained within distinct membrane-bound organelles called protein bodies (Figure 6.5). In the seeds of legumes and other dicotyledonous groups where protein is stored in the cotyledons, the membrane of the protein body is retained during seed dormancy so that the protein is surrounded by the membrane during its mobilisation on seed germination as well as during storage. This is in contrast to the situation in the starch endosperm of wheat where the membrane breaks down during the final stages of grain development, so that the membrane-free protein bodies are liberated and stick to the surface of the starch granules in the endosperm and to one another. Thus in the mature wheat endosperm, storage protein is present not so much in the form of discrete bodies, but more as a matrix within which the starch granules are embedded. This difference is consistent with the different roles of cotyledon and endosperm during germination: the cotyledon becomes a fully active functional orga-nelle, while the endosperm serves as a passive source of nutrients for the embryo. The difference may also be related to the need to maintain the storage proteins out of solution. The globulins of the legumes, being soluble in salt solutions, might well dissolve in the cell cytosol. Hence they may need to be maintained within an organelle where the salt concentration can be controlled at a low level. The prolamins of the cereal endosperm are insoluble in salt solutions and therefore would not require such protection.

BIOSYNTHESIS

Amino acid synthesis

The storage proteins are made from the same 20 amino acids as are the other proteins of the plant, although there may be a predominance of certain amino acids in the storage proteins, such as the proline and glutamine of the prolamins of cereal. For the synthesis of amino acids a source of carbon skeletons, and of nitrogen in the form of amino groups, are both required. In cereals most of the nitrogen stored in the grain is obtained from the vegetative parts, either derived from the remobilisation of protein or from the foliar reduction of nitrate to the amide group of glutamine. Also, in other protein-yielding seeds a large proportion of the stored nitrogen is derived from the vegetative parts of the plant. Thus nitrogen arrives to the seed mainly in the phloem and in the form of asparagine and glutamine. Phloem is also the source of the sucrose from which the carbon skeletons are formed.

The origins of the carbon skeletons are shown in Figure 8.5. Most amino acids are derived from phosphoglycerate, phosphoenolpyruvate and pyruvate tapped from the glycolytic pathway, and 2-oxoglutarate and oxaloacetate from the TCA cycle. Additional carbon skeletons are derived from erythrose 4-phosphate and ribose 5-phosphate.

In the developing seed, the amide nitrogen of glutamine and asparagine is

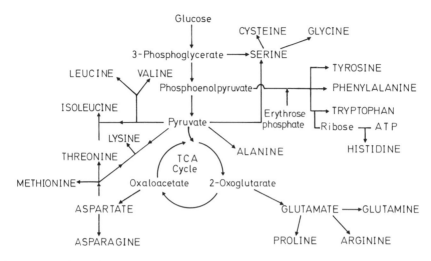

Fig. 8.5 The origins of the carbon skeletons for amino acid biosynthesis. (Redrawn from Miflin, 1980. Reproduced by kind permission of the author and Academic Press.)

released for incorporation into the synthesis of the other amino acids, generally by a transamination reaction from glutamate. The mechanisms by which the syntheses of amino acids are regulated has attracted attention from investigators because of the potential for the partial release of that control so as to increase the production of nutritionally limiting amino acids. The regulation of amino acid synthesis can be illustrated by the synthesis of lysine, methionine and threonine, which can all be present at undesirably low levels in storage proteins. The starting point for the biosynthesis for all three amino acids is aspartate (Figure 8.6). The first enzyme, aspartate kinase is sensitive to inhibition by lysine and by threonine. In some plants there are two isoenzymes, one sensitive to lysine, the other to threonine. Methionine does not inhibit aspartate kinase directly, but a derivate, S-adenosylmethionine can enhance the sensitivity to lysine. Thus the levels of all three amino acids influence the activity of aspartate kinase. Dihydropicolinic acid synthase catalyses the first reaction after the pathway branches towards lysine. This enzyme is inhibited by lysine. Similarly the first enzyme after the pathway has diverged towards methionine and threonine, homoserine dehydrogenase, is inhibited by threonine but not by lysine. It is not, however, inhibited by methionine (Figure 8.6).

The recognition of these feedback mechanisms as the means by which the syntheses of amino acids like lysine are regulated, suggested a potential means of relieving their deficiency in cereal and legume seeds: the selection of mutants in which control is relaxed, and which consequently overproduce the limiting amino acid. Overproducing mutants have been derived from cultured cells, or in the case of cereals (where regeneration from culture is difficult)

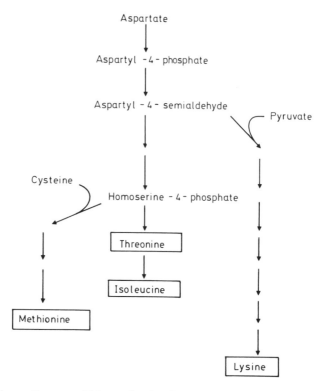

Fig. 8.6 The pathways of biosynthesis of methionine, lysine and threonine. Multiple arrows indicate individual enzymatic steps in the sequences.

embryos have been used. However, there are a number of considerations that weaken the basis of this approach towards the generation of crop plants with improved amino acid balance. First there is compartmentation. Amino acids for the most part are made in plastids, stored in vacuoles and used in the cytoplasm at the rough endoplasmic reticulum. Thus *in vivo* they probably do not accumulate in the compartment in which they are made, but are translocated away, and the significance of any feedback inhibition is correspondingly diminished. Second, the regulation of a particular enzyme activity is likely to change with the developmental stage of the plant and with the type of tissue, so that selection for over-production at one stage would not necessarily provide the desired over-production at another stage.

Protein synthesis

Synthesis of the storage proteins is directed by the specific mRNA on polyribosomes attached to the endoplasmic reticulum. During synthesis the protein is transported across the membrane of the endoplasmic reticulum.

Near the N-terminus of the polypeptide there is a sequence of about 20 amino acids containing a high proportion of hydrophobic residues. This signal sequence acts to direct the growing polypeptide chain towards the endoplasmic reticulum membrane where it is recognised by a specific channel protein. During the synthesis of the polypeptide the entire chain is drawn through the membrane into the lumen of the endoplasmic reticulum. Within the lumen the signal sequence is cleaved from the nascent chain by a specific protease. As the polypeptides aggregate within the lumen, it swells and takes on the character of the protein body. Clear evidence for this pattern of development was obtained by Larkins & Hurkman (1978) in developing grains of maize. In the early stages in the development of the endosperm, the protein bodies of maize grains can be seen under the electron microscope to be closely associated with the rough endoplasmic reticulum. Ribosomes are attached to the membrane of the protein bodies as well as to the endoplasmic reticulum. The protein bodies can be separated from fragments of the rough endoplasmic reticulum by sucrose gradient centrifugation of homogenised developing maize grains. It is then possible to show that the mRNA of polyribosomes of both rough endoplasmic reticulum and protein bodies synthesise identical prolamins, and that the marker enzyme of endoplasmic reticulum is also represented in the protein body membrane. Thus in its development the protein body retains features of the endoplasmic reticulum from which it originates.

In developing protein-rich seeds large amounts of rough endoplasmic reticulum can be seen to develop at the stage of the most rapid protein deposition. This rough endoplasmic reticulum is easily extracted from the immature seeds and is found to be enriched in mRNA sequences specific for storage proteins. Moreover this mRNA can direct the synthesis of storage proteins *in vitro*, if supplied with rough endoplasmic reticulum together with the amino acids, tRNAs and the other factors necessary for protein synthesis.

Protein synthesis is maximal during the latter phase of seed development, when cell expansion has taken over from cell division. Immediately preceding the onset of protein accumulation there is a drastic change in the population of mRNA molecules, so that diversity is lost and a limited number of mRNA sequences become very abundant. Figure 8.7 shows how in soybean cotyledons the mRNA for a single protein class, the soybean legumin glycinin, comes to represent 10 per cent of the total mRNA during the maximum period of protein accumulation. This is equivalent to over 30 000 molecules of glycinin mRNA per cell.

Once polymerisation is complete the storage proteins undergo a series of specific post-translational changes, which are characteristic for the different types of storage protein. In pea, legumin and vicilin appear to be present in the same protein bodies, but the post-translational modifications undergone by each differs. These changes have been followed by pulse-labelling the protein with radioactive amino acids (Chrispeels *et al.*, 1982a,b). These studies show that the legumin is made first as a single polypeptide chain of M_r 60 000 containing both polypeptide subunits of M_r 40 000 and 20 000, which

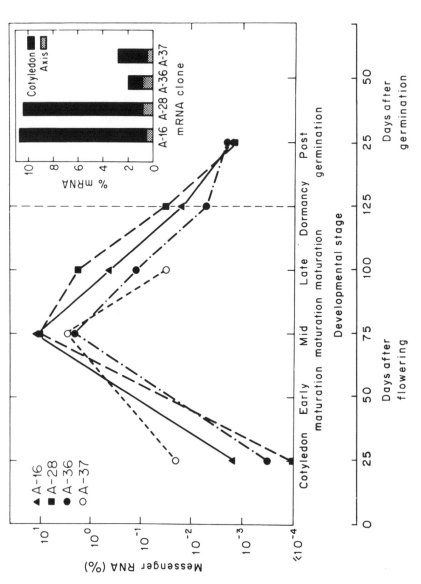

Fig. 8.7 Changes in the level of soybean glycinin mRNA during seed development and germination. (From Goldberg *et al.*, 1981. Reproduced by kind permission of the authors and of Academic Press.)

are subsequently separated by a post-translational proteolytic cleavage, which probably takes place within the protein body. After cleavage the two subunits remain bonded together by disulphide bonds. Similar post-translational modifications probably take place in the synthesis of the legumin-like proteins of broad bean, soybean, and oat and in the synthesis of rice glutelin.

The mature vicilin of pea contains subunits with a M_r of about 75 000 and 50 000 as well as a range of smaller polypeptides, but all the peptides of M_r < 50 000 have sequences related to parts of the polypeptides of M_r 50 000. Not unexpectedly therefore these smaller peptides have been found to be initially synthesised as peptides of M_r 50 000, which are subsequently cleaved to form the smaller polypeptides (Chrispeels *et al.*, 1982a,b). The mannose- and N-acetylglucosamine-rich oligosaccharide side chains of vicilin are attached covalently to asparagine residues. Pulse-chase labelling experiments demonstrate that the incorporation of ^{14}C-labelled glucosamine occurs in both the rough endoplasmic reticulum and in a subsequent stage in the Golgi apparatus.

The zein accumulated by maize is not a homogenous compound and the question arises of how the different types of zeins are ordered and accumulated within the protein bodies. Lending & Larkins (1989) have used antibodies specific for α-, β- and γ-zeins and immunogold labelling of tissue sections to locate these zeins cytologically during development. Their findings have enabled them to propose a model for the developmental sequence followed in the formation of protein bodies in maize (Figure 8.8). At the earliest stage the protein bodies contained β- and γ-zeins, but no α-zein. As the cells of the endosperm matured, the protein bodies increased in size and the α-zein content increased. Initially the α-zein appeared as isolated locules, but as α-zein began to predominate, the locules fused to form a central core with the β- and γ-zeins largely confined to a peripheral zone (Figure 8.8), with small patches or strands of γ-zein embedded within the central region.

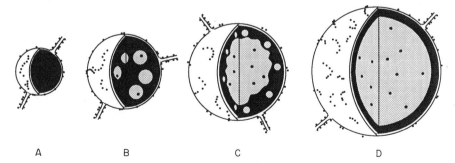

A   B   C   D

Fig. 8.8 Development of protein bodies in maize endosperm. The protein body is surrounded by endoplasmic reticulum with ribosomes attached. Dark shading represents β and γ zeins; lighter shading, α zein. (From Lending & Larkins, 1989. Figure courtesy of the authors. Reproduced by kind permission of the American Society of Plant Physiologists.)

BIOTECHNOLOGICAL DEVELOPMENTS

The improvement of the yield and quality of protein in legumes and cereals has long been the aim of plant breeders. Quite what is intended by quality depends very much on the particular crop in question. In legumes one of the principal aims in the improvement of protein quality has been the removal of toxins that accompany the protein in the seed, for example the neurotoxic non-protein amino acids of grass pea that cause the crippling disease known as lathyrism. In oilseeds the quality of the oil rather than that of the protein has been of over-riding importance, but in oilseed rape, when the breeders reduced the glucosinolates they lowered the toxicity of the protein-rich cake which remains after the oil has been extracted and in this way improved protein quality. For malting barley a high quality variety will have grain with a low protein content. There are a number of reasons for this. Low protein levels are associated with the desirable high carbohydrate contents; the storage protein can envelope the starch grains and thus interfere with sugar production and fermentation; and protein combines with polyphenols to give an insoluble precipitate which leads to cloudiness in the beer.

However, there are two broad categories of improvement which relate directly to the biosynthesis and molecular structure of the protein itself. First, we have the improvement in the nutritional value of the protein. Second, we have the improvement in the physical properties of the protein for particular biotechnological applications, particularly as the properties relate to the bread-making quality of wheat varieties. In both of these cases analysis of the molecular structure of the major storage proteins has revealed a biochemical basis for the desired qualities.

Improvement in the nutritional quality of plant proteins

The nutritional value of plant proteins is determined largely by their amino acid compositions in comparison to the requirement of man and his non-ruminant livestock, the pigs and poultry, for certain amino acids which have to be supplied in the food. These essential amino acids account for half of the 20 amino acids which make up proteins (Table 8.1).

Ruminant animals can obtain a supply of essential amino acids from microorganisms in their rumen. However, this source of amino acids is not available for non-ruminant livestock, which are often intensively reared and bred for rapid rates of growth, and thus have a particularly high demand for the essential amino acids. In maize and wheat, two of the most important sources of livestock-feed in the world, there is a serious deficiency of lysine, with tryptophan as the second limiting amino acid after lysine. Generally legume proteins have adequate amounts of lysine, with soybean being particularly rich in lysine, but in legume proteins the S-containing amino acid methionine becomes limiting.

When a limiting amino acid fails to meet the nutritional requirements of the animal, the other amino acids are used as a source of energy rather than in the synthesis of protein, and the excess nitrogen is excreted. In practice the deficiency of one protein source is overcome by mixing cereal and legume feeds, by adding fishmeal, or by supplementing the feed with amino acids of synthetic origin. However, overcoming the basic imbalance in the amino acid composition of the plant source by supplementation is expensive. One approach to overcoming the problem would be to transform genetically livestock so that the animals themselves could synthesise the deficient amino acids (Rees *et al.*, 1990). The transgenic animal, having received the appropriate genes from a source such as a microorganism, would then be independent of its feed for that amino acid, and consequently more efficient in the conversion of a cereal-based feed. This is not a solution available for all livestock, nor of course to the problem of human nutrition. A more widely applicable solution would be to alter the amino acid balance of the proteins accumulated by the crop plants.

In cereals a number of mutant lines with increased levels of lysine have already been discovered. These high lysine lines have been found in maize, sorghum and barley (Doll, 1984). In all lines examined the increased lysine content has been shown to be due largely to a reduction in the prolamin content. In some barley lines there was a compensating increase in the synthesis of globulins which are relatively rich in lysine, but in general terms the mutant lines could more accurately be described as low-prolamin rather than high-lysine. Moreover the decreased prolamin content was accompanied by a decreased synthesis of starch in the endosperm, and this caused the grains to shrivel and grain yields to fall. The mutant lines were therefore unattractive despite their improved nutritional value.

Since the time of this work on the low-prolamin, high-lysine lines, the application of molecular biology to the study of the major storage proteins has provided us with a greater understanding of storage protein structure and regulation. In particular it is now realised that the primary sequences of the prolamins are dominated by repeated sequences rich in proline, glutamine and hydrophobic amino acids (Figure 8.4). The basic amino acids such as lysine are absent from these sequences.

Based on our present knowledge of the molecular biology of the seed proteins, there are broadly two approaches to the improvement in the amino acid balance of the seed proteins. First, the particular storage protein that is deficient in an amino acid could be substituted by protein of higher quality. This could be accomplished by enhancing the concentration of a protein already present, but poorly represented. One technique here would be to substitute a regulatory sequence of the gene for the presently under-represented protein with the regulatory sequence of the highly expressed protein of inferior quality. Alternatively, the gene for the desired protein could be transferred from one crop to another, for example transfer from soybean to maize of the lysine-rich glycinin.

The second approach is to alter the amino acid composition of the deficient protein. The work of Wallace *et al.* (1988) illustrates well this approach. They showed that when the mRNA of the α-zein of maize (M_r, 19 000) was enriched for lysine codons, and injected into frog oocytes for *in vivo* translation, the resulting proteins could be deposited in the oocytes as stable, densely packed membrane-bound protein bodies. Three types of modification were carried out: neutral amino acids in several positions were substituted by lysine; lysine-rich peptides were inserted into the zein molecule; and a large stretch of an unrelated hydrophilic protein was inserted into the N-terminal region. After the first two modifications the modified zein aggregated as normal into a densely packed protein body. Only the last modification created a sufficiently drastic alteration in zein structure to prevent the normal deposition pattern. But assembly into a dense protein body is only one of the requirements of a storage protein. For example, the protein also needs to take up an appropriate structural conformation during grain drying and rehydration. Nevertheless, the ability of these cloned, lysine-enriched zeins to assemble into a protein body indicates that the zein molecule may be sufficiently tolerant of lysine insertions for an engineered lysine-rich zein to be a practical possibility.

Bread-making quality and the proteins of wheat

To make bread, flour is mixed with water and yeast to form a dough, which is then rested to allow fermentation by the yeast. The carbon dioxide so produced is trapped within the dough, giving it a light porous crumb structure, which is fixed when the bread is baked. The ability of a flour to produce bread depends on the visco-elastic proteinaceous mass called gluten, which is what remains when the starch is washed from dough. Gluten extracted from wheat can itself be considered as a crop product: in 1987 48 000 tonnes of gluten were used in the UK (Shewry & Tatham, 1989) to fortify flours for making speciality breads such as wholemeal.

Glutens can also be obtained from barley and rye, but not from maize. However, only wheat flour possesses the appropriate properties to give the porous, spongy quality of bread. Good quality bread is produced when the gluten gives the dough an appropriate balance of two physical properties: viscous flow and elasticity. In the wheat varieties grown in Europe, bread-making quality is limited by poor elasticity rather than by poor viscosity. The viscosity of gluten is determined by the gliadins, while the elasticity is determined by the HMW subunits of the glutenin polymers (Table 8.5). Breadmaking quality is strongly correlated with the presence of the HMW subunits, and different cultivars of bread wheat which show good or poor breadmaking quality have different allelic forms of the HMW subunits. Analysis of the secondary structure of the HMW subunits reveals that while the flanking N-terminal and C-terminal regions are globular, the central region constructed of repetitive blocks of amino acids (Figure 8.4) has an unusual β-spiral structure (Shewry & Tatham, 1989). In this conformation the protein chain regularly folds back on itself in a series of reverse turns (Figure 8.9). It has been proposed that this β-spiral conformation provides the elasticity of the HMW subunits. The helix of stacked β turns would be

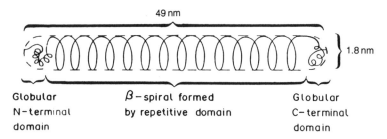

Fig. 8.9 Structural model of the HMW subunit of wheat gluten responsible for bread making quality. (From Shewry & Tatham, 1989. Reproduced by kind permission of the authors and of Pergamon Press.)

stabilised by hydrogen bonding and hydrophobic interactions. When stresses are applied during the working of the dough the molecules would be stretched and the hydrogen bonds and other interactions would be broken to be reformed when the stress was relieved, hence giving the dough its required elasticity. This proposal is supported by a comparison of the likely secondary structure of HMW subunits with other elastic proteins, such as the elastin of mammalian connective tissues.

Identification of the molecular basis of the elasticity of glutenin opens the way to the improvement of the breadmaking quality of wheat varieties, as exemplified by the work of Flavell *et al.* (1989). They sequenced two allelic genes which coded for HMW subunits which conferred different baking qualities to the dough. The non-repetitive sequences at the N- and C-termini were identical, including the positions of the cysteine groups which form the disulphide bridges that link the subunits into aggregates. However, the differences in amino acid sequences gave different predicted β turn structures, with the superior allele possessing more repeat units than the inferior allele, that is, with longer regions of predicted β turn structure. This finding allows us to predict the amino acid modifications that might be introduced into the superior gluten HMW subunit to enhance further its contribution to dough quality.

FURTHER READING

Derbyshire, E., Wright, D.J. & Boulter, D. (1976). Legumin and vicilin, storage proteins of legume seeds, *Phytochemistry* **15**, 3–24.

Doll, H. (1984). Nutritional aspects of cereal proteins and approaches to overcome their deficiencies, *Phil. Trans. Roy. Soc., London B* **304**, 373–380.

Gatehouse, J.A., Croy, R.R.D. & Boulter, D. (1985). The synthesis and structure of pea storage proteins, *CRC Crit. Rev. Plant Sci.* **1**, 287–314.

Higgins, T.J.V. (1984). Synthesis and regulation of major proteins in seeds, *Annu. Rev. Plant Physiol.* **35**, 191–221.

Ingversen, J. (1983). The molecular biology of storage protein synthesis in maize and barley endosperm, in *Seed Proteins*, Eds Daussant, J., Mossé, J. & Vaughan, J., London, Academic Press, pp. 193–204.

Kreis, M. & Shewry, P.R. (1989). Unusual features of cereal seed protein structure and evolution, *BioEssays* 10, 201–207.

Larkins, B.A., Pederson, K., Marks, D.M. & Wilson, D.R. (1984). The zein proteins of maize endosperm. *Trends Biochem. Sci.* 9, 306–308.

Miflin, B.J. (1980). Nitrogen metabolism and amino acid biosynthesis in crop plants, in *The Biology of Crop Productivity*, Ed Carlson, P.S., New York, Academic Press, pp. 255–296.

Miflin, B.J., Field, J.M. & Shewry, P.R. (1983). Cereal storage proteins and their effect on technological properties, in *Seed Proteins*, Eds Daussant, J., Mossé, J. & Vaughan, J., London, Academic Press, pp. 255–319.

Payne, P.I. (1983). Breeding for protein quantity and protein quality in seed crops, in *Seed Proteins*, Eds Daussant, J., Mossé, J. & Vaughan, J., London, Academic Press, pp. 223–253.

Pernollet, J.-C. & Mossé, J. (1983). Structure and location of legume and cereal seed storage proteins, in *Seed Proteins*, Eds Daussant, J., Mossé, J. & Vaughan, J., London, Academic Press, pp. 155–191.

Rees, W.D., Flint, H.J. & Fuller, M.F. (1990). A molecular biological approach to reducing dietary amino acid needs, *Bio/Technology*, 8, 629–633.

Shewry, P.R. (1991). Barley seed storage proteins—structure, synthesis and deposition in *Nitrogen Metabolism of Plants*, Eds Mengel, K. & Pilbeam, P.J., Oxford, Oxford University Press, 201–227.

Shewry, P.R. & Miflin, B.J. (1985). Seed storage proteins of economically important cereals, in *Advances in Cereal Science and Technology*, Ed Pomeranz, Y., Vol VII, Saint Paul Minnesota, American Association of Cereal Chemists Inc., pp. 1–83.

Shewry, P.R., Miflin, B.J. & Kasarda, D.D. (1984). The structural and evolutionary relationships of the prolamin storage proteins of barley, rye and wheat, *Phil, Trans. Roy. Soc. London B* 304, 297–308.

Shewry, P.R. & Tatham, A.S. (1989). New light on an old technology: the structure of wheat gluten and its role in breadmaking, *Outlook on Agriculture* 18, 65–71.

Shewry, P.R. & Tatham, A.S. (1990). The prolamin storage proteins of cereal seeds: structure and evolution, *Biochem. J.* 267, 1–12.

Spencer, D. (1984). The physiological role of storage proteins in seeds, *Phil, Trans. Roy. Soc., London B* 304, 275–285.

Thompson, G.A. & Larkins, B.A. (1989). Structural elements regulating zein gene expression, *BioEssays* 10, 108–112.

Wright, D.J. & Bumstead, M.R. (1984). Legume proteins in food technology, *Phil. Trans. Roy. Soc. London B* 304, 381–393.

ADDITIONAL REFERENCES

Borroto, K. & Dure III, L. (1987). The globulin seed storage proteins of flowering plants are derived from two ancestral genes, *Plant Mol. Biol.* **8**, 113–131.

Brinegar, A.C. & Peterson, D.M. (1982). Separation and characterization of oat globulin polypeptides *Arch. Biochem. Biophys.* **219**, 71–79.

Chrispeels, M.J., Higgins, T.J.V., Craig, S. & Spencer, D. (1982a). Role of the endoplasmic reticulum in the synthesis of reserve proteins and the kinetics of their transport to protein bodies in developing pea cotyledons, *J. Cell Sci.* **93**, 5–14.

Chrispeels, M.J., Higgins, T.J.V. & Spencer, D. (1982b). Assembly of storage protein oligomers in the endoplasmic reticulum and processing of the polypeptides in the protein bodies of developing pea cotyledons, *J. Cell. Biol.* **93**, 306–313.

Flavell, R.B., Goldsbrough, A.P., Robert, L.S., Schnick, D. & Thompson, R.D. (1989). Genetic variation in wheat HMW glutenin subunits and the molecular basis of bread-making quality, *Bio/technology* **7**, 1281–1285.

Goldberg, R.B., Hoschek, G., Ditta, G.S. & Breidenbach, R.W. (1981). Developmental regulations of cloned superabundant embryo mRNAs in soybean, *Dev. Biol.* **83**, 218–231.

Larkins, B.A. & Hurkman, W.J. (1978). Synthesis and deposition of zein in protein bodies of maize endosperm, *Plant Physiol.* **62**, 256–263.

Lending, C.R. & Larkins, B.A. (1989). Changes in the zein composition of protein bodies during maize endosperm development, *Plant Cell* **1**, 1011–1023.

Wallace, J.C., Galili, G., Kawata, E.E. Cuellar, R.E. Shotwell, M.A. & Larkins, B.A. (1988). Aggregation of lysine-containing zeins into protein bodies in *Xenopus* oocytes, *Science* **240**, 662–664.

Glossary of Plant Names

Barley	*Hordeum vulgare* L.
Borage	*Borago officinalis* L.
Cape marigold	*Dimorphotheca pluvialis* Moench
Cassava, tapioca plant	*Manihot esculenta* Crantz
Castor, castor bean	*Ricinus communis* L.
Chickpea	*Cicer arietinum* L.
Chicory	*Cichorium intybus* L.
Coconut	*Cocos nucifera* L.
Cotton	*Gossypium hirsutum* L.
Crambe	*Crambe abyssinica* Hochst. ex. Fries
Cuphea	*Cuphea lutea* Rose
Currants	*Ribes* spp L.
Dahlia	*Dahlia pinnata* Cav
Elecampane	*Inula helenium* L.
Evening primrose	*Oenothera biennis* L.
Field bean, broad bean	*Vicia faba* L.
Flax, linseed	*Linum usitatissimum* L.
French bean	*Phaseolus vulgaris* L.
Garden nasturtium	*Tropaeolum majus* L.
Goldenrod	*Solidago altissima* L. (*S. rupestris* Raf.)
Grape	*Vitis vinifera* L.
Grass pea	*Lathyrus sativum* L.
Groundnut, peanut	*Arachis hypogaea* L.
Guayule	*Parthenium argentatum* Gray.
Hemp	*Cannabis sativa* L.
India rubber tree	*Ficus elastica* Roxb.
Jerusalem artichoke	*Helianthus tuberosus* L.

Jojoba	*Simmondsia chinensis* (Link) Schneider
Jute	*Corchorus capsularis* L.
Lentils	*Lens culinaris* Medic.
Lesquerella	*Lesquerella fendleri* (A. Gray) S.Wats.
Lupin	*Lupinus albus* L.
Maize, corn	*Zea mays* L.
Meadowfoam	*Limnanthes alba* Hartw.
Milkweed	*Euphorbia lagascae* Spreng.
Millets	*Pennisetum americanum* Auth. (*P. glaucum* R. Br), *Eleusine coracana* Gaertn., *Panicum miliaceum* L.
Oats	*Avena sativa* L.
Oil palm	*Elaeis guineensis* Jacq.
Oilseed rape	*Brassica napus* L.
Olive	*Olea europaea* L.
Pea	*Pisum sativum* L.
Potato	*Solanum tuberosum* L.
Rice	*Oryza sativa* L.
Rubber euphorbia	*Euphorbia Tirucalli* L.
Rubber tree	*Hevea brasiliensis* Muell, Arg.
Rye	*Secale cereale* L.
Safflower	*Carthamus tinctorius* L.
Sea Island cotton	*Gossypium barbadense* L.
Sesame	*Sesamum indicum* L.
Sisal	*Agave sisalana* Perrine
Sorghum	*Sorghum vulgare* Pers.
Soybean	*Glycine max* Merr.
Spinach	*Spinacia oleracea* L.
Sugar beet	*Beta vulgaris* L.
Sugar cane	*Saccharum officinarum* L.
Sugar maple	*Acer saccharum* Marsh.
Sugar palms	*Arenga saccharifera* Labill., *Borassus flabellifer* L.
Sunflower	*Helianthus annuus* L.
Sweet potato	*Ipomoea batatas* Lam.
Taro, cocoyam	*Colocasia esculenta* Schott
Timothy grass	*Phleum pratense* L.
Tomato	*Lycopersicon esculentum* Mill.
Wheat	*Triticum aestivum* L.
White mustard	*Sinapis alba* L.

Index

Index compiled by Annette Musker

ACS-7849   10/9/95